T0076379

Contents

List of Figures

List of Figures

List of Tables

List of Contributors

Xira Ruiz-Campillo is a professor of Environmental Regimes and International Relations at the Complutense University of Madrid. She holds a PhD in International Relations from the same university. She has worked at the Spanish Ministry of Foreign Affairs, the Spanish Ministry of Defence and at the United Nations High Commissioner for Refugees' (UNHCR) Headquarters in Geneva. She has several publications on conflict prevention and crisis management. Her current research interests are related to climate change and sustainable policies in the European Union as well as the role of cities at the international level.

María del Pilar Bueno is a researcher at the National Council for Scientific and Technical Research of Argentina CONICET, teacher and professor at several universities including Universidad Nacional de Rosario, Universidad Nacional de Entre Rios, Universidad Nacional de La Plata and Universidad de San Andrés, among others. Climate change negotiator and advisor of the Argentinean Ministry of Foreign Affairs since 2013; leading negotiator of the G77 and China on Adaptation from 2016 to 2018; co-chair of the Adaptation Committee since 2018 and member of the Task Force of Displacement at the UNFCCC from 2017 to 2018.

María Dolores Sánchez Galera LLB (Honours) Glasgow University, LLM, PhD (Scuola Superiore Sant'Anna, Pisa). She is currently a CONEX-Marie Curie Fellow at Carlos III University (Madrid), member of the Urban Land Planning and Environmental Law Pascual Madoz Institute of Carlos III University and member of the Sustainable Development Solutions Network (SDSN) (Spain). She has been a legal officer for IDLO and has published extensively in international journals on different environmental and energy law topics. She has taught undergraduate- and postgraduate-level courses on Environmental Law and Sustainability in different European Universities.

Kattya Cascante Hernández is a professor in Complutense University of Madrid at the International Relations and Global History Department on Faculty of Political Science and Sociology. She holds a PhD in International Relations by the Complutense University of Madrid. She has extensive experience as a consultant for the evaluation of projects led by the European Union (EU) and other international agencies or programs responsible for development countries. Part of her relevant publications focuses on development and sustainable cooperation system as well as on the speculation on the food system and food crisis.

Elena Bulmer Santana holds a PhD from the Complutense University of Madrid; PhD Candidate in Project Management applied to Stakeholder Management at the University of Manchester; PMP® by P.M.I.®; PRINCE-2; Fundraising Certificate. She has 15 years of international experience in the management of environmental conservation and biodiversity projects. She presently teaches at various business schools and academic institutions such as E.A.E. Business School-Madrid Campus.

María Isabel Nieto is a professor of European Union Policy and International Relations at the Complutense University of Madrid. Her research interests focus on European Neighbourhood Policy, regional policy and international relations. Apart from her academic experience, she has a long professional experience on European politics and international relations, having worked as an advisor in European Affairs for local and regional governments in Spain since 2003.

Rosa Giles-Carnero is a professor of Public International Law at the University of Huelva and member of the research group Public Law and Governance. She holds a PhD in Public International Law and has published on international and EU environmental law, peace and security, and gender studies. Currently, her main research interest is related to international action on climate change, and environmental governance in a global context.

Ekaterina Domorenok is associate professor of Political Science at the University of Padua. Her research interests mainly concern policy design, implementation and learning in multi-level settings, with particular regard to European Union policies for climate, sustainability, environment and cohesion. She has taken part in several European research and cooperation projects dealing with regional and local development, environmental governance, climate policies and participatory forms of policy-making at the local level. She is the author of four monographs and has published a number of articles on the aforementioned topics.

Introduction

It is evident that climate change has impacted our lives and our environment. Without any doubt, it was developed countries the most responsible for accelerating that change in our climate although developing countries are increasingly greatly contributing to it. It is our responsibility as human beings to leave future generations with a better world or, at least, not one worse. We are not the owners, but only the casual dwellers of a planet that has given us the opportunity to live and to enjoy a breathtaking environmental wealth that we must preserve. It is the duty of all countries and citizens to take care of what we have been given for the years we are here, and it is our responsibility to analyse the best way of doing that.

This book is a collective effort made by eight women concerned and united by the research on climate change, sustainability and cooperation policies within the European Union. This work tries to shed light on the evolution the European Union has experienced since 1992 in its fight against climate change in the search of a more sustainable Europe. Without wanting to be exhaustive, the book aims to depict the evolution of the European Union's policies on climate, cooperation, sustainability, security, economy and energy with the final aim of being a trustworthy and reliable actor at the international level. The impact that international agreements on climate, cooperation and sustainability have had on European policies is analysed in different chapters.

It was not intentional that all the authors were women, but probably what is offered here is a unique perspective on how we see the European Union is managing and weathering the most challenging threat of the 21st century.

Chapter 1 analyses the transformation that the European Union (EU) has undergone to incorporate international climate agreements. Agreements at the international level such as the Kyoto or the Paris Agreement have led the EU to adopt binding internal targets to reduce its greenhouse gas emissions and increase its share of renewable energy that have resulted in the promotion of a circular economy and the change of the energy system within the European borders. Both a circular economy and renewable energy are considered strategic areas to set the EU into a sustainable path of growth. Although the promotion of sound and ambitious climate and sustainable policies at the international level has been part of the European identity for long, the truth is that its credibility depends on the achievement of EU targets–nothing easy to achieve.

María del Pilar Bueno analyses the changes experienced in the EU leadership in climate change diplomacy, in particular, at the United Nations Framework Convention on Climate Change (UNFCCC) from Copenhagen Summit in 2009 to Katowice Conference in 2018. This period is portrayed by an adjustment in the leadership of the EU that can be considered in isolation after the withdrawal of the United States from the Kyoto Protocol. However, Copenhagen illustrated the beginning of a shared leadership with the United States and China that continues to generate great challenges to the EU regionally and internationally. Considering the vast trajectory of leadership literacy applied to climate change negotiations, the chapter underscores four features of EU leadership at the international climate change arena including its vision, internal coherence, the degree of achievement of its objectives and proposals and the flexibility to adapt to changing contexts.

In her chapter, María Dolores Sánchez Galera presents an overview of how the evolution of EU environmental law and governance has taken the lead towards the emergence of a unique EU sustainability model with its imperfections and challenges, but with greedy ambitions within the global scenario. All in all, the EU sustainability model is, of course, a dynamic process resembling the European integration process itself. A complex process full of changes triggered by globalisation and the global governance processes born, especially after the end of the Cold War. This analysis aims to deliver greater understanding of the actual legal commitment of the EU to implement sustainable development laws, principles and policies that can only be understood through the lenses of a multidisciplinary framework that could help us to unpack the complex global paradigm of sustainable development.

Kattya Cascante Hernández analyses how the commitments adopted in the Agenda 2030 have impacted European cooperation and the development cooperation policies of the EU and how the EU has been present in the International Development Cooperation policy. Europe's support of the International Development Cooperation policy has overcome the weight of the financial crisis and the isolationist behaviour of other relevant actors, although the EU may be wasting the opportunity created by the distancing of China and the USA from the implementation of the Sustainable Development Goals (SDGs) Agenda. All that political performance should have been reflected in the improvement of the European leadership to advance, without gaps and with less rhetoric, a form of governance based on sustainable development, responsible and coherent with the commitments acquired and, most importantly, the initiative of constructing common global goods to address climate change.

Elena Bulmer Santana describes the strive of the EU for a low-carbon, resource-efficient and competitive economy through the development of a circular economy. The step from a linear to a circular economy can be understood as a tool to reduce emissions and set the EU towards a more sustainable Europe. In addition, this new economy will protect businesses from the risk of scarcity and will develop new opportunities and jobs at all levels of the value chain. The current implementation of a circular economy nevertheless is still in its initial stages and has to cope with different barriers such as the actual production model, the resource consumption processes or societal behaviours.

In her chapter, María Isabel Nieto seeks to clarify the progress made to date on the EU's energy policy, especially with the entry into force of the Lisbon Treaty. Energy, innovation and sustainability are examined in the chapter. Although the challenges are huge and steps have been taken to establish a common energy policy and an integrated climate and energy policy of the EU since 2014, results have proven mixed. The chapter examines the Juncker Commission's transition to cleaner energy and many of the policies and instruments underway: a whole battery of programmes, normative acts and measures aimed at achieving an integrated energy market, the interconnection of energy networks and the proposals in the domain of energy efficiency and renewable energy sources, among others.

Rosa Giles-Carnero, through a securitisation lens, analyses how the EU is confronting the effects of climate change. As she points out, using a securitisation approach for the analysis of governance provides a new opportunity to incentivise the adoption of actions to mitigate and to adapt to climate change. The chapter looks into the concept of climate security and stresses how climate security instruments in the EU play a significant contribution to the EU's credibility in international climate negotiations and how the relationship between climate change and security has been reinforced in EU documents as the EU has aimed to increase its role in climate negotiations.

Ekaterina Domorenok focuses on how enhancing polycentric policy making and strengthening the role of sub-state authorities have been among the EU's main endeavours in the effort to comply with its international commitments in the field of climate policies. The chapter explores the experience of the EU Covenant of Mayors programme that was launched by the European Commission in 2008 in order to enhance and support the local action for climate change across EU countries. Besides presenting an overview of the programme's evolution over time, the analysis focuses on the functioning and the main policy tools incorporated therein, bringing also some evidence on how and why cities join this initiative, and to what extent they exploit the opportunities provided by the Covenant of Mayors membership.

CHAPTER 1

Leadership in the European Union: Climate, Energy and Economy

Xira Ruiz-Campillo

The status quo is not an option. Countries should act together to protect their citizens against climate change.
–European Commission (2018b)

Over the years, an enormous amount of research has been devoted to analyzing the role of the European Union (EU) in climate negotiations (Zito, 2005; Falkner, 2007; Afionis and Stringer, 2012). However, not much has been written about how international obligations on the matter and the EU's assumed role as a normative power in this regard have impacted its objectives in relation to the reduction of greenhouse gas emissions (GHGs), nor how the path toward their achievement has transformed the EU's own development, energy and economic models. For example, obligations to reduce GHG as part of the Kyoto Protocol necessarily lead to changes in industry, making it more efficient and less polluting. The objectives adopted at the European level in relation to the increasing use of renewable energy also lead to a change in the current energy model, from a model in which the use of fossil energy was predominant to one in which renewable energy not only contributes to reducing GHG, but also increases the energy security of European states. The promotion of a circular economy and the initiative of creating an Energy Union are both examples of the transformations that the EU is experiencing. Both initiatives will help improve the environment by reducing GHG emissions and by increasing the production of renewable energy, while both, as a whole, will help the EU achieve its international and domestic commitments, thus consolidating its leadership in global environmental governance. What the chapter examines is thus how the EU has been transformed as a result of the agreements achieved in the international domain and its position as a global leader in environmental issues. Concretely, the chapter will analyze how international agreements and the EU's leadership have fostered the adoption of domestic goals and, as a result, how the EU is reinforcing its international leadership and promoting the adoption of robust international agreements.

The EU has demonstrated its commitment to climate change over the past 25 years, as well as its ability to mobilize other countries in ensuring that the reduction of GHG and the promotion of renewable energy are on the agenda of the main international forums. For example, the EU played a key role in the success of the Paris Agreement through the promotion of its climate diplomacy and that of groups such as the High Ambition Coalition in which, together with African, Caribbean and Pacific countries, it strove to achieve support for the 2015 agreement. This leadership has led the EU to proactively search for international agreements as well as internal commitments leading to a transformation of the European energy and economic models. As a result, the EU is strengthening not only its leadership in climate negotiations, but also, for example, in the renewable energy sector, which was one of the political priorities of the Juncker Commission (European Commission, 2016).

The EU is not just a leader in climate change action, but also in other environmental areas. The Seventh Environment Action Programme (2013), for instance, established nature and biodiversity environment health and quality of life and natural resources and wastes as priority areas, all of which are important to ensure the EU's long-term prosperity (European Union, 2013). The object of this chapter, however, is more concrete. It focuses on how the adoption of international objectives in the field of climate and energy, particularly to reduce GHGs, has increased the EU's leadership in climate negotiations as well as transformed it internally, impacting in areas such as the economy and the energy models. Achieving binding objectives has led to a transformation of the industry, the creation of jobs and a transformation in the fields of technological innovation and renewable energies, among others.

The EU began in 2007 a path toward the objective of net-zero GHGs with the aim of creating a more sustainable and powerful Europe in the global environmental governance field by 2050. A non-emissions Europe that was to be reached in a series of intermediate stages in which the EU would adopt binding goals—so far the 2020 and 2030 goals—together with an intense international leadership where the EU would strive for ambitious global decisions. More than a decade after 2007, it is fair to say that the EU has adopted significant decisions with an impact on the economy and the use of energy in order to meet the 2020 goals, the first binding objectives established in the field of climate and energy. To meet the 2030 goals and to reduce emissions by 80%–95% by 2050, however, the EU will need to radically transform the energy system, buildings, transport, the land and agricultural sectors and modernize the industrial fabric and cities of Europe (European Commission, 2018a).

The chapter begins by analyzing the EU as a normative power with a strategic approach in climate negotiations that seeks to extend its own vision on how to fight climate change, an area where it has found a genuine sense of responsibility and support from European citizens. The emergence and evolution of EU leadership in climate negations is examined to then analyze the impact that international agreements have had in the adoption of the 2020 goals. Part three and four, respectively, analyze the adoption of the 2030 goals and how this has transformed the energy and economy models of the EU, particularly through the promotion of a European Energy Union and a circular economy.

1.1 The EU as a Normative Power in Environmental Negotiations

The EU has historically been considered a leader by example in climate negotiations (Zito, 2005; Falkner, 2007; Afionis and Stringer, 2012), having adopted high targets in relation to the reduction of GHGs, approving a vast number of legally binding norms to protect the environment and investing in renewable energy, among other things.

The internal transformation of the EU as a result of its fight against climate change can be analyzed as part of this normative power, following the work of a great number of academics (Manners, 2002; Whitman, 2011). Manners stated that 'civilian' power involved three key dimensions: the centrality of economic power to achieve national goals; the primacy of diplomatic cooperation to solve international problems and the willingness to use legally-binding supranational institutions to achieve international progress (Manners, 2002). These three dimensions in both the internal and international domains are fundamental to understand the transformation experienced by the EU's environmental, economic and energy models. In all these areas, economic growth through more sustainable means is one of the foundations of the transformation taking place therein, while diplomacy has been key to spread the EU interests and has been fundamental to the signing of instruments such as the Kyoto Protocol and the Paris Agreement, as will be explained below. Indeed, the transformation the EU is experiencing cannot be understood without its understanding as a normative actor *determined to promote economic and social progress for their peoples, taking into account the principle of sustainable development and within the context of the accomplishment of the internal market and of reinforced cohesion and environmental protection,* as stated in the Amsterdam and subsequent treaties.

In line with Manners' definition of 'normative power Europe', academics have analyzed the EU as a model (or rule-setter), as a player or rule-negotiator shaping world events and negotiations; as an innovator; as a market and as an instrument or rule-facilitator for global governance (Barbé et al., 2015; Zito, 2005). The role of the EU as a normative power has also been contested, however, and part of the literature has cautioned against the counterproductive consequences of the absence of a more strategic approach in international relations in a world where transnational challenges are gaining increasing importance and where non-Western powers are becoming more powerful in an increasingly globalized world (Howorth, 2010; Mayer, 2008; Barbé et al., 2015). Analyzing the role of the EU in the environmental arena as that of a normative power, however, can go hand in hand with the adoption of a more calculated strategic approach. As Howorth points out, strategic alliances will be key in achieving objectives in the areas of trade, energy, development and the environment among others (2010). Indeed, the EU has already proved that it is capable of dropping its preferred options in climate negotiations in order to avoid losing its leadership, thus following a clear strategic approach. The EU wants to exert its influence in the world and has shown in negotiations to have specific goals, as well as the utmost importance that it ascribes to values and norms, which is the main difference between civilian power and soft imperialism (Hettne and Söderbaum, 2005). Thus, the EU can be seen as a normative power, pursuing a norm-driven foreign policy, which first stems from the values it promotes internally, such as human rights, the rule of law, democracy and sustainable development (Hettne and Söderbaum, 2005). In the past decades, the EU has been the key driving force behind the harmonization of environmental standards among the EU's Member States (Falkner, 2007), such model also having an effect on third countries. At the same time, the EU has shown an incredible capacity to adapt to new international climate contexts—as it demonstrated with the reemergence in 2009 of the United States in climate negotiations—without ever forgetting its own normative model, however, which is nothing else than the imposition of its own vision on how climate change should be faced. It is important to always analyze the normative power argument pragmatically, therefore, taking into account that when acting as a global environmental leader, the EU is also defending its interests, acting 'to change norms in the international system' and 'making its external relations informed by, and conditional on, a catalogue of norms' (Manners, 2002; 2011). Thus, as Falkner (2007) and Afionis and Stringer (2012) argue, a critical reading of the normative power argument is necessary, which adds an interest-based perspective to international environmental negotiations.

1.1.1 *The Emergence of the EU's Environmental Leadership*

For an actor who has been widely criticized for not being a leader in other areas such as security (Howorth, 2010; Mayer, 2008), the environmental arena has been key for the EU to acquire a more prominent and recognized role in the globalized world. The EU's environmental leadership is indisputable, though such role has had ups and downs since the 1990s. The EU chose to have a notable role in the environmental arena in an effort to consolidate its identity as a normative or civilian power (Kelemen, 2010), precisely because the environmental arena was one in which the EU would not be eclipsed by other international actors. The EU is able to maintain its leadership in the environmental arena through two main strategies exerted in the international and domestic levels: first, in the international domain, as Kelemen (2010) and others point out, the EU tries to impose its own vision on environmental regulations, so that they do not go against its own interests and somehow reflect the EU's own domestic mode. Second, to reinforce its environmental leadership internationally, the EU has to lead by example, and therefore adopt ambitious domestic environmental policies in order to fulfill its international commitments, from the binding Kyoto Protocol to the more flexible Paris Agreement.

The EU has gradually strengthened its international profile regarding environmental issues since the beginning of the 1990s, having adopted a progressive approach to sustainability (Falkner, 2007). In *The Environmental Imperative Declaration*, the European Council (1990) recognized its 'special responsibility for the environment' regarding both its citizens and the rest of the world, accepting to play a leading role in promoting concerted and effective action at global level; an endeavor that it began by requesting the Commission to have a more active role in relation to the environment, conducting regular reviews and publishing reports. A year after, and a few months before the 1992 Rio Summit, the Commission issued its first strategy to limit carbon dioxide (CO_2) emissions and to improve energy efficiency. The strategy included a package of proposals including research and development (R&D) programs, the promotion of renewable energy and fiscal and voluntary measures among others (Commission of the European Communities, 1991). The document prompted the members of the European Community to 'provide both leadership and example to developed and developing countries in relation to the protection of the environment and the sustainable use of natural resources' in order to preserve the benefit of present and future generations (Commission of the European Communities, 1991). The publication of initiatives, of policies or communications by European institutions months before an international conference on the environment or on development has been a regular modus operandi of the EU, as a way to influence the international negotiations. In this way, the EU tries to gain power internally through the adoption of a common position that will then be defended internationally by Member States and EU representatives alike. Thus, adopting decisions that imply the reduction of GHG or the protection of the environment shows an internal compromise that reinforces European leadership in international conferences. In this regard, the support of Member States to stringent environmental policies has been key to reinforce the EU's leadership at the international level, albeit in some cases reluctantly and not as ambitiously as it would have been desired.

The signing of the 1992 United Nations Framework Convention on Climate Change (UNFCCC) is widely recognized as the first success of European countries in climate negotiations (Torney, 2013; Kelemen, 2010; Oberthür and Pallemaerts, 2010). Taking advantage of its increasing international role in environmental affairs and of the weakened US leadership at the time (Falkner, 2007; Ruiz-Campillo, 2017), the EU has managed to acquire a prominent role in international environmental negotiations ever since, backed by a comprehensive body of environmental norms and standards adopted at the EU level. European environmental leadership since the 1990s can be explained as the result of a combination of domestic politics (political forces within Europe both at the national and EU level) and international regulatory competition (Kelemen, 2010). At the international level, the role of the EU as a normative power is clearly recognized in the environmental arena, where it has played a significant role and managed to spread its own vision on how to manage the environment. Such role, however, cannot be fully explained without considering the domestic domain, where a mix of pressure from environmental groups and northern European countries coexist with a longer history of environmental regulations and with the European Commission, which proposes environmental regulations in an attempt to keep up with its environmental prominent role in a globalized world. Another factor behind the EU's environmental leadership is the constant support of European citizens for environmental action. Citizens' support has grown since 1992, when 72% of respondents supported the protection of the environment as an area where the EU should take the lead (Eurobarometer, 2008); a support that increased to 87% in 2017, together with 94% of respondents who think that protecting the environment is important (Eurobarometer, 2017). The environment is not only defended in Europe through politics—European green parties have had representatives in the Parliament since the creation of the Rainbow Group in 1984—but also through an intensive work of green lobbies.

The area that most interest arouses among lobbies in the EU is, according to the EU Transparency Register (as of January 2019), the environment, followed by R&D, business and industry, energy and climate action.

During the last two decades and with the exception of a decline between 2009 and 2012, the EU has led the environmental arena without competitors due to the absence of the United States and China. The Commission, with an exclusive right of initiative in environmental issues and under exclusive competence when operating externally, has managed to take the EU's leadership to its highest (Vogler and Stephan, 2007), proposing an extensive number of ambitious initiatives going from the creation of a European energy market to the ban of single-use plastics. At the domestic level, the accession of Sweden, Austria and Finland to the EU in 1995 strengthened the bloc of pro-environment states (Kelemen, 2010), together with the changes in the decision-making process brought about by the Maastricht Treaty, which imposed qualified majority voting instead of unanimity and reinforced environmental issues in its title XVI, the role of the EU in environmental issues grew stronger. Indeed, majority voting is seen by the EU as key to upgrade the level of ambition in a particular area where a high impact is intended, as has been the case with environmental issues (European Commission, 2018b). Domestic European policies during the 1990s, however, were not impressive and a gap between international promises and domestic implementation became the Achilles' heel of the EU's environmental leadership during such period (Oberthür and Roche Kelly, 2008; Kulovesi, 2012). Yet this credibility gap began to close after the adoption of the Kyoto Protocol in 1997 and the Marrakech Accords in 2001 (Oberthür and Roche Kelly, 2008), as their entry into force led European leaders to adopt a vast number of initiatives in order to reduce GHG, taking the EU to one of its most prominent positions of leadership in international relations.

Indeed, the EU assumed a clear leading role in the negotiations that led to the signing of the Kyoto Protocol in 1997. Already during the Rio Conference of 1992, European states expressed their desire for a binding agreement to reduce GHG emissions; however, it was not until Kyoto that the idea crystallized (Chasek et al., 2014). Although the EU was committed to undertake measures to reduce GHG regardless of the outcome of Kyoto, it was important for its leadership to have a wide support from developed countries—the ones who would adopt binding commitments—to the Kyoto Protocol, a treaty that the EU had carefully crafted. It was not until the EU convinced Russia to join Kyoto in return for supporting its bid for membership in the World Trade Organization, that Kyoto was finally implemented (Chasek et al., 2014; Schunz, 2011). The consolidation of European leadership in Kyoto—but also in Rio—coincided with the total absence of the United States (Vogler and Stephan, 2007) and with a reinforced diplomatic activity demonstrating other actors a European commitment to reducing emissions with the ultimate goal of obtaining support to the Kyoto Protocol (Schunz, 2011). Through the framework provided by Kyoto/UNFCCC, the EU has been able to encourage developing countries to join the effort to combat climate change and to adopt EU standards and technologies when doing so (Kelemen, 2010), and at the same time, the international role played by the EU in the environmental governance has been reinforced by initiatives and policies undertaken at the domestic level.

During Kyoto's first commitment period (2008–2012), in the face of the indifference and even the rejection of big polluters (United States and China) to join the fight against climate change, it was the EU that exercised such leadership in climate negotiations. Leaving aside the ability of Europe to make other countries converge around a common position in the fight against climate change during this period, it is necessary to highlight how international commitments had an influence in the decisions adopted by the EU to reduce emissions and how such decisions helped transform the European industry itself as it will be explained below. Keeping global environmental leadership, it is argued here, has demanded the EU to comply with international agreements and thus transform its economy and energy market, which is a clear evidence of 'leading by example'. An example of this was the proposal, within its Kyoto compromises, of reducing 8% its industrial GHG, the deepest emission cuts and the highest reduction targets among the major industrialized countries (Oberthür and Roche Kelly, 2008).

1.1.2 *European Leadership in Climate Negotiations Since 2009*

Diplomacy has been used as a tool to take the EU's environmental leadership further in the last years, having become pivotal to gain support for ambitious climate goals that would pave the way to a lower emissions Europe (Schunz, 2011; Torney, 2013). Since 2007, and despite the continued adoption of actions conducive to European growth and sustainable development, the EU began to lose its prominent position in international climate negotiations (Ruiz-Campillo, 2017). This responds to several factors: on the one hand, the global economic crisis affected European countries in a different way, which emphasized the divisions among the EU's Member States, as they had different visions on how to face the economic downturn. This contributed to the withering of the EU's international role as it was no longer able to speak with a single voice in many international meetings. A clear example of this is the Conference of the Parties (COPs) in Warsaw in 2013, in which the host country, Poland, was on the verge of making the summit fail. On the other hand, the arrival of Barack Obama to the US government meant that the EU began to be overshadowed by the new leader's eagerness for leadership in the area of the environment. A good example of the rapid loss of European prominence in climate negotiations is found at the Copenhagen Summit of 2009. Although the EU came up with an innovative and courageous proposal against climate change, it suffered a strong humiliation when relegated and isolated from the negotiations led by the United States and the BASIC countries, which were to draft the Copenhagen Accord. The Agreement, which was only taken note of because it was negotiated outside the UNFCCC process, was novel and laid the foundations for the negotiations in subsequent COPs, where the climate architecture until then designed by the EU and without the participation of powers such as United States or China would be completely inverted.

The EU's interest in and commitment to the environment has a long history, unlike its ability to influence climate change negotiations. Although the EU has led climate negotiations for years, it is also true that such leadership may have been achieved thanks to the indifference that successive US presidents have shown with respect to the environment, thus allowing the EU to fill the position of leadership in its absence (Ruiz-Campillo, 2017). The return of the United States in 2009 entailed the loss of leadership in negotiations for the EU and the restructuring of the regime that the EU had shaped until then. From this date, the climate regime started moving from a top-down approach to a bottom-up one led by the United States: an approach in which each State contributes to the fight against climate change based on its own commitments and possibilities.[a] The crafting of the Copenhagen Accord with the absence of the EU was a wake-up call for the EU and its until then unquestionable leadership, which forced it to look for new ways in which to assert its leadership in climate negotiations at the international level. As a result, the EU put a new emphasis on climate diplomacy, which, while not new for the EU, had not been used to its fullest potential in relation to the environmental agenda.

An additional sign of the EU loss of leadership can be found in the second Kyoto's commitment period (2013–2020), in which only the EU together with other European countries and Australia had binding targets after the withdrawal of Canada and the decision of Russia to not participate in this period of commitment. Extending Kyoto was, however, important for the EU's leadership, as it would become the only instrument to fight climate change in place until the Paris Agreement entered into force.

Between 2012 and 2015, the EU doubled its diplomatic efforts in order to reach an agreement to be signed in Paris, an example of this was the decision of making the European External Affairs Service to lead the Green Diplomacy Network. In the international domain, it was important that the EU employed all its diplomatic tools and it introduced climate change in all the international and bilateral meetings. At the domestic level, it was important to adopt clear ambitious targets that would reinforce its diplomatic efforts to acquire a leading role in the negotiations. Thus, in 2014, with the Paris Agreement on the horizon, the European Council adopted the 2030 climate and energy action framework, which proposed a 40% reduction of GHG emissions and a 27% increase in the use of renewable energy and energy efficiency by 2030. The objective of reducing GHG emissions by 40% was indeed the EU's contribution to the Paris Agreement in 2015. In its Nationally Determined Contributions (NDC)—contributions in the fight against climate change that each country offers to the

[a]For a deeper analysis on the role of the EU in climate negotiations, see María del Pilar Bueno's chapter in this book.

international community, but whose non-compliance can only lead to little more than a possible loss of international prestige—the EU also mentioned the possibility of reducing European emissions by 80% by 2050 should a favorable international context be in place, an objective already proposed by the Commission as early as 2011. Thus, the 2030 reduction targets were established in parallel to the negotiations of the Paris Agreement, independent from it but mandatory for EU Member States. The framework adopted was the most ambitious among the major players in the environmental arena and its achievement will entail further transformations in the EU. As such, they must be analyzed as part of the leading role that the EU wanted to exert in the upcoming Paris Agreement and in the context of the 2030 Agenda for Sustainable Development.

At the end of 2015, the Paris Agreement, the first international agreement to fight climate change in which all the countries adopted GHG reduction commitments was signed. If the Kyoto Protocol turned into an instrument that basically forced European states to reduce their emissions, the Paris Agreement has the virtue of having agreed that all states, regardless of their level of development, reduce their GHGs through the NDCs. The signing of the Paris Agreement was at the same time important for the adoption of goals at the European level, as the success of these climate negotiations can be seen as a necessary condition for ambitious European unilateral goals to be adopted (Fisher and Geden, 2013).

The Paris Agreement entered into force on 4 November 2016, not even a year after it was signed and only two days before the US elections that gave the victory to Donald Trump; surely as a strategy to avoid the United State's absence from the Paris Agreement from the very beginning and as a symbol of the legacy of Obama era. Although this was possible thanks to the EU's partial ratifications (not all European countries had ratified the Paris Agreement by that time), the truth is that China, the United States and India had ratified the agreement before the EU, reducing the leadership the EU should have played. Given that the United States, with the new administration, is no longer interested in leading the environmental arena, it is again the EU that should deliver and take advantage of the momentum created in Paris, especially since China seems to be in the path toward a more sustainable development.

1.2 From the International to the Internal Domain: The Impact of the Kyoto Protocol and the Paris Agreement on the EU

As seen above, the adoption of legally-binding policies within European countries has been key to strengthen European international leadership and its compliance with international agreements. To meet the commitments set out in the Kyoto Protocol, the EU adopted the Emissions Trading System (ETS) in 2003. Moreover, to prove its concern with the environment domestically as well as internationally, EU Member States adopted the 2020 targets in 2007. In order to meet all these self-imposed binding objectives, the EU has had to transform the European industry, energy, transport and market.

1.2.1 *The Kyoto Protocol and the ETS System*

The Kyoto Protocol (1997) sets the pace for some initiatives that would emerge and develop within the EU, as the emissions trading scheme of 2003 or the effort sharing decision (ESD) adopted in 2009 (Decision nº 406/2009/EC). In the year 2000, the European Commission established the first European Climate Change Program (ECCP) (2000–2004) with the aim of helping European states reach the Kyoto protocol commitments demanding them to cut their combined GHG emissions to 8% below the 1990 level by 2012. The first ECCP established a number of working groups to examine how areas such as agriculture, research, energy, industry or transport could reduce their GHG emissions, whereas the second ECCP (2005–2010) explored cost-effective options for achieving significant GHG emissions while simultaneously increasing growth and job creation. As a result, the EU emissions trading scheme system (ETS), currently covering around 45% of the EU's GHG emissions, was created through Directive 2003/87/EC, but only began to be implemented in 2005, once Kyoto entered into force.

The EU ETS has had an impact on heavy energy-using industry (power generators and industrial plants) and airlines companies (since 2013) operating in the 31 countries in which it applies (all 28 EU countries plus Iceland, Liechtenstein and Norway). The EU ETS forces major industrial installations in five industrial sectors (combustion—including electricity production, district heating, cogeneration and refineries—metal—including iron and steel—cement, glass, ceramic and paper-board productions) to reduce their CO_2 emissions (Delbosc and Perthuis, 2009). Through the EU ETS, both industrial companies and States have to assume binding objectives, which have forced the latter to change their national regulations in turn forcing industrial companies to reduce their emissions. The EU ETS has evolved over the years in different phases (2005–2007; 2013–2020 and 2021–2030) incorporating more sectors and gases to be reduced, developing an EU-wide cap on emissions (instead of national caps) and helping the industrial sector in their carbon transitions. Putting a price on carbon[b] contributed to reduce emissions by 11.8% between 2008 and 2012 in European territory; a reduction that has been achieved in part due to the switch from coal to natural gas, to the employment of more efficient coal and to the improvement of CO_2 efficiency (Delbosc and Perthuis, 2009).

To support the reduction of emissions at EU level, the EU is also guiding its industrial transformation through different funding mechanisms, such as NER 300, established by article 10a(8) of the EU Emissions Trading Directive and currently one of the world's largest funding programs, which seeks to boost the deployment of innovative low-carbon technologies (such as carbon capture and storage (CCS) and innovative renewable energy) as well as to stimulate the creation of jobs in relation to such technologies within the EU. This is an example of how a commitment adopted at the international level (Kyoto Protocol) impacts the European level (the creation of the EU ETS) and in turn the national level (through the creation of obligations regarding the regulation of the industry in order to reduce its GHG emissions), while imposing a new direction on the industrial and scientific sectors through the prioritization of investments and research on low-carbon technology and the promotion of green jobs.

Covering a population of more than 512 million people, 28 countries and 45% of the European emissions, the EU ETS is an initiative that has

[b]The penalty for non-compliance was €40 per ton in the first commitment period. In the second phase, the penalty increased to €100 per ton. The aviation sector was brought into the UE ETS on 1 January 2002 although its application was suspended for that year (European Commission, 2019).

clearly empowered the EU in the environmental arena. The EU ETS was the first market of its kind to be created and has contributed to at least three things: first, it has helped the EU meet its commitments in the Kyoto protocol (it achieved an overall cut of 11.7% in the first commitment period, without counting LULUCF reductions and international credits); second, it has served as a model for other markets that emerged after—New Zealand in 2008, China in 2013 or Korea in 2015 (Tuerk and Zelljadt, 2016)—and third, both things have contributed to strengthen the EU's leadership in relation to the environment. The EU ETS has been considered the boldest (and yet successful) experiment to date in the use of offsets (Ellerman et al., 2016) and will continue helping the EU to reduce its GHG, to guarantee a carbon price and to transform the energy system.

1.2.2 *The 2020 Package: Binding Domestic Reductions to Lead International Negotiations*

In March 2007, EU leaders agreed on an ambitious long-term strategy to reduce GHG and to increase the use of renewable energy by 2020. The agreement can be seen as a way of promoting the measures agreed upon in Kyoto with regard to enhancing energy efficiency, increasing new and renewable forms of energy and encouraging reforms in relevant sectors in order to reduce GHG emissions as set in article 2 of the Protocol. But, most importantly, the 2007 agreement can be seen as a tool to increase the EU's leadership before the 2009 Copenhagen Conference, where a post-Kyoto agreement was expected to be signed. The rationale behind the 2020 framework adopted by the EU to face climate change is clear in both the documents of the European Council and those of the Commission, all of which reflect how the EU acts bearing in mind both the domestic and international dimensions: at the European level, the cost of fighting climate change will be higher the longer states wait, so the sooner European states act, the cheaper and easier adapting to climate change will be. At the international level, the EU is aware that the earlier it moves, the greater opportunity it will have to spread its skills and technology to other states.

The 2007 European Council's agreement would mark the real beginning of a number of EU climate policies that would transform first the energy model and then the economy in European territory. The agreement set binding legislation that committed European countries to (1) reduce GHGs by 20% in the so-called diffuse sectors (all those not included in the Kyoto protocol and that represent 55% of the total emissions of the EU, such as residential, commercial, transport, agriculture, livestock, etc.); (2) to increase the consumption of renewable energy by 20% and (3) to increase energy efficiency also by 20%, all by 2020. According to the president of the European Commission at the time, José Manuel Barroso, the package was said to be 'the most ambitious package ever agreed to by any Commission or any group of countries on energy security and climate protection' (New Scientist, 2007). The Presidency Conclusions of March 2007 offered a clear support to the Commission's energy and climate change package proposal at the domestic level and to the 2007 scientific report of the Intergovernmental Panel on Climate Change, underlining the importance of limiting the increase of temperature to no more than 2°C above pre-industrial levels and, to do so, to adopt an integrated approach to climate and energy policy (European Council, 2007).

Given that energy production and use are the main sources for GHG in the EU, an integrated approach to climate and energy policy was identified as key and states agreed to develop three main objectives within the Energy Policy for Europe (European Council, 2007): increasing the security of supply (which would entail transforming the energy market creating a single energy market); ensuring the availability of affordable energy and the competitiveness of European economies (where the circular economy will have a central role) and promoting environmental sustainability and combating climate change (accelerating the shift to low carbon energy and accepting a binding target of a 20% GHG reduction). In addition, the Presidency Conclusions mentioned the need for developed countries to collectively reduce their GHG emissions in the order of 30% by 2020 compared to 1990, and by 60%–80% by 2050 compared to 1990 (European Council, 2007). These decisions adopted by EU leaders enhanced the 2001/77/EC and 2003/30/EC directives on renewable energy that already stated the need to promote renewable energy sources to contribute to environmental protection, but did not include any yet binding targets for Member States.

As indicated above, agreeing on a solid internal position on environmental policies can be seen as a way of strengthening European leadership in climate negotiations as well as a way to enhance the necessary conditions for reaching a new global agreement in 2009 to follow on from the Kyoto Protocol's first commitment after 2012 (Commission of the European Communities, 2007). Indeed, in the second commitment period Parties committed to a global reduction of at least 18% below 1990 levels and the EU, together with Iceland, committed to a 20% reduction target as had been previously established with the adoption of the 2020 targets.

In order to meet the targets established in relation to emissions and in renewable energy and efficiency by 2020, the Commission proposed a set of regulations in 2008 known as the EU Climate and Energy Package that were enacted in legislation in 2009. One of the most important regulations in the package was the ESD 406/2009/EC (ESD) that entered into force in 2009 and established a binding GHG emission target for the period 2013–2020 in areas not covered by the EU ETS, such as buildings, waste, agriculture or transport. As in the Kyoto Protocol, the 2020 goals also took into account the different levels of development of European countries. Thus, reduction goals went from the highest 20% of Ireland or Denmark, to the increase allowance of 20% for Bulgaria. As a whole, the EU would reduce by 20% its emissions by 2020.

An update to the ESD was approved in 2018 through regulation 2018/842, which established binding reductions for the period 2021–2030, with a target of at least 40% domestic reductions by 2030 compared to 1990 levels, which would contribute to meet EU commitments under the Paris Agreement. The ESD is part of the Energy Union strategy and the EU's implementation of the Paris Agreement, and includes provisions that give Member States some flexibility in order to meet their targets—such as the borrowing and transfer of unused emissions or the use of GHG reduction projects in third countries—while if states do not meet the goals, they risk receiving a penalty (European Union, 2016). Contrary to the EU ETS, which is regulated at EU level, the ESD gives states flexibility to act in whatever areas they want (from promoting public transport to promoting climate-friendly farming practices) as long as GHG reductions are pursued.

The 2020 targets are a clear example of how the EU has adopted initiatives to comply with the spirit of the UNFCCC and the Kyoto Protocol's goal of reducing global emissions, and within a time that led the EU to adopt more ambitious targets, later defended at the international level, in order to show leadership. At the same time, however, international agreements such as Paris have had an impact on the EU itself, as seen with the 2018 update of the ESD. To reach the targets on GHG emissions and on renewable energy, a strong commitment at all levels is needed. With the aim of engaging European cities and towns in reaching the EU climate and energy targets, the European Commission also launched in 2008 the Covenant of Mayors initiative, transformed over the years in order to adapt to the consecutive targets European countries have adopted.[c] As a result of the mobilization of these sub-state actors, cities and towns of all sizes across the EU have been empowered in the fight against climate change, and have collectively have reduced 23% of their GHG emissions through the reduction of emissions from buildings, transport, heating, etc. (Joint Research Centre, 2016). In 2015, the EU announced the partnership with the Compact of Mayors—initiative launched in 2014 by the the UN Secretary General Ban Ki-moon and former New York City Mayor Michael Bloomberg, the UN Special Envoy for Cities and Climate Change—resulting in the Global Covenant of Mayors for Climate and Energy in 2017. European cities are the most numerous in this global initiative, which replicates the EU model in the establishment of determined targets for signatories, showing again the EU's leadership in the climate regime.

1.3 The Transformation of the EU's Energy Model

The transformation of the energy model in Europe has always had a strong link with environmental protection from an institutional point of view (Thieffry, 2016). Title XXI of the Treaty on the Functioning of the European Union (TFEU) (2007) already highlighted in article 194 the need to preserve and improve the environment, and states as policy aims the energy security, the promotion of energy efficiency and energy saving and the development of renewable energy. For the EU, energy has a key role to play in the transition toward a net-zero GHG economy as it is responsible for more than 75% of the EU's GHG emissions (European Commission,

[c]See E. Domonerok's chapter in this book for a wider understanding of the mobilization of cities in the fight against climate change.

2018a). Therefore, complying with international climate agreements such as Kyoto or Paris necessarily entails the transformation of its energy system by increasing its energy efficiency and the share of renewable energy. However, one of the main difficulties when adopting measures in this sector lies on Art. 192 of the TFEU, which states that the EU's energy policy cannot affect Member State's choice between energy sources nor the general structure of its energy supply unless it counts with the Council's unanimous agreement.

The path toward an increase in renewable energy production as a means to reducing CO_2 emissions is not new in Europe. Already in 1997, the European Commission published a White Paper on renewable energy proposing to reach a share of renewable energy of 12% by 2010, as a means to reduce CO_2 emissions, to decrease energy dependence and to develop national industries and create jobs (European Commission, 1997). Directive 2001/77/EC included indicative targets for each Member State, although the lack of progress toward achieving such 2010 targets prompted European states to adopt new and binding targets in 2007.

In its 2007 Presidency Conclusions, the Council established five priority actions in the energy sector: the creation of an internal market for gas and electricity to enhance competitiveness, benefit consumers and reduce energy dependence; to increase energy supply by increasing cooperation among states; the development of a common approach to external energy policy by intensifying cooperation with third parties; the increase by 20% of energy efficiency and the share of renewable energies; and the strengthening of technologies in the energy sector (European Council, 2007). The establishment of these priority areas led to the adoption of Directive 2009/28/EC to promote the use of at least 20% of the energy from renewable sources by 2020. The directive sets binding goals for Member States, ranging from the highest, a 49% increase for Sweden (which started from almost 40%) to the lowest, 10% for Malta (whose effort would actually be higher as it started from 0%). In addition, the directive added a binding goal for all states of a 10% target for renewable sources in transport. This directive was considered to be a key to economic growth, the creation of jobs and to improve the energy security of the EU, and is the main instrument to ensure that Member States comply with the objective of increasing renewable production by 20% in 2020 and to reduce GHG (e.g., the increase in renewable energy prevented the emission of 362 Mt in 2013 and 380 Mt of CO_2 in 2014) (EEA, 2016).[d]

In February 2011, the European Council reaffirmed the objectives endorsed by world leaders in the Copenhagen and the Cancun Agreements of reducing 80%–95% GHG by 2050 (European Commission, 2011). Only a couple of weeks after, the European Commission proposed a number of initiatives destined to set the EU in the path of reducing those emissions by mid-21st century. The *Roadmap to a competitive low carbon economy in 2050* (2011) acknowledged the importance of adopting ambitious domestic climate change policies for the EU's global environmental leadership and how not adopting a bold plan to promote a low carbon economy would lead the EU to lose ground in major manufacturing sectors (European Commission, 2011). It is worth noticing that it was in the 2009 Copenhagen conference when the EU suffered a loss of leadership, which acted as a trigger that led the EU to adopt a stronger internal position and to use climate diplomacy as a tool to lead the negotiations that would result in the adoption of the Paris Agreement, as mentioned earlier.

In 2014, before the Paris summit, and again as a means of stepping up international commitment to the fight against climate change, European States agreed on a framework that would increase the share of renewable energy in at least 27% by 2030. The lack of ambition on the part of States, the evidence that not enough was being done and the recognition that energy was a growing political and economic challenge led the Commission to propose in 2017 a revision of the framework on Renewable Sources of Energy Directive (2009) in order to accelerate the transition toward becoming the world's number one region on renewable energies, proposing that the target of renewable energies reached at least 30% by 2030. This was also as a response to the realization that the proposed commitments were not enough to limit the average increase in temperature to 2°C as agreed in Paris.

European targets and the Paris Agreement will be difficult to meet without a successful transition to a clean energy system (European

[d]For more detailed information on the European energy model transformation, see the chapter of María Isabel Nieto in this book.

Commission, 2016). This is why one of the main instruments to meet the 2030 goals proposed in 2014 was the Energy Union. The 2016 *Clean Energy* package puts emphasis on achieving global leadership in renewable energies and energy efficiency, for which it proposes a reviewed target of 30% energy efficiency. The Commission's package led the European Council to adopt new binding targets during 2018: a 32% in renewable energy (through European Directive 2018/2001; OJEU, 2018a) and an energy efficiency target of at least 32.5% to be reached at the European level (European Directive 2018/2002; OJEU, 2018a). These targets, together with the cross-border electricity interconnection of 15% by 2030, will transform the European energy system into a more sustainable one; it will improve industrial competitiveness, boost growth and jobs and reduce GHG emissions while improving the quality of air, or at least those are the EU expectations. To ensure that the EU's 2030 energy and climate targets are achieved, Member States agreed to create by the end of 2018 a new governance system for the energy union that aims to transform Europe's energy system in order to preserve, protect and improve the environment, to promote a prudent use of natural resources, to offer consumers secure, sustainable and affordable energy and to foster research and innovation by means of attracting investment (OJEU, 2018b).

Through the adoption of successive initiatives, during the last decade, the EU has managed to increase its share of renewable energies from 10.4% in 2007 to 17% in 2017, even though the EU's forecasts indicate that, with the current policies, only a 24.3% of renewable energy consumption will be achieved by 2030, far from the 32% established as a binding target, which means that significant reforms at all levels will be necessary. This renders more urgent the materialization of a single energy market that ensures price competitiveness and makes investments more attractive. Renewable energy is a sector in which the EU has already experienced a drop in investment of around 60% with respect to 2011 and whose leadership is beginning to be transferred to Asia. If one of the EU's objectives is to become a world leader in renewable energies, both in terms of consumption and in the commercialization of technology (European Commission, 2016), increasing the production of renewable energy is a necessary requisite, and at the same time a fundamental part of its economic development in the medium and long term.

1.4 The Transformation of the European Economy

Economy has always been the engine of the EU and the reduction of GHGs is necessarily linked to economic activities. The EU is experiencing a profound economic transformation and it is bound to seek that others in the international domain follow suit. One of the milestones of this transformation was the publication of the *Europe 2020 Strategy* (European Commission, 2010), adopted in a difficult context for the EU. In addition to the humiliation it experienced in the 2009 Copenhagen negotiations, when the United States sought to play the leader's card, the EU was experiencing an acute economic crisis that was capable of derailing the efforts made to curb gas emissions. The *Europe 2020 Strategy* emerged as a response to the European crisis in 2010, with the aim of boosting employment, improving education in younger generations, investing more in R&D and meeting the 20-20-20 targets. In the *Strategy*, there is a clear link between reaching such goals and fighting climate change. For instance, the development of technologies such as carbon capture and sequestration, and boosting efficient and renewable energy is seen as a way to both soothe the economic crisis while reducing GHG emissions (European Commission, 2010).

The period of recession beginning in 2008 allowed the EU to make another economic reconversion, just as it did in the early 1990s, though this time the investment in renewable energies and the improvement of energy efficiency were the driving forces behind this new transformation (Ruiz-Campillo, 2017). These transformations allowed the economy to continue growing in absolute terms despite the fall in emissions: between 1990 and 2011, EU gross domestic product (GDP) grew 45%, whereas emissions decreased an 18.3%; and between 2010 and 2011, GDP increased 1.4%, whereas GHG declined 3.3%. These facts have been widely used by European diplomats to convince other states that combating climate change does not necessarily mean reducing economic competitiveness, and thus to promote an economy based on low emissions at the international level. This low-emissions economy, promoted through a circular economic model, can be analyzed as another instrument through which the EU has exported its policies both directly (as with the memorandum of understanding signed between the EU and China to promote a circular economy in July 2018) and indirectly, through, for example, the high-level political and business meetings organized by the Directorate-General for the Environment of the European Commission with the aim of promoting sustainable and resource-efficient policies within third countries, NGOs and companies.

The development of a new type of economy was outlined in the *Europe 2020 Strategy*, which did not explicitly refer to the term 'circular economy', but did make references to the development of 'a smart, sustainable and inclusive economy delivering high levels of employment, productivity and social cohesion'.

The essence of a circular economy is the idea that raw materials remain within the life cycle as long as possible and, when a product or raw material cannot be reused or recycled anymore, that waste is as little harmful as possible to the environment.[e] This is what authors William McDonough and Michael Braungart (2002) call the 'from cradle to cradle' process, in which the production model would resemble the life cycles of nature and as such, would produce no waste, as everything has a purpose in the circle of nature. The concept of circular economy combines old and well-established notions of resource efficiency while making explicit the economic aspect of saving resources and the potential gains it accrues (Milios, 2018). Circular economy is clearly linked to the reduction of GHG and the increase of renewable energy as GHG emissions are emitted in all stages of the product life cycle: in extraction, production, consumption and waste management (Behrens, 2016), meaning that a deep transformation of the economy (the use of raw materials, improved resource efficiency and greater recycling and re-use) will be needed. A more circular economy is seen in some reports as a great contributor to deep cuts of emissions as a result of the better use and reuse of materials in heavy industries as well as in the production of goods consumed in the EU (Deloitte, 2016; Sitra et al., 2018).

Making the economy circular implies a transformation in the processes of production and consumption at European level and as such, should go in hand with the support of both the industrial sector—responsible for integrating into their creation processes the ideas that are behind a circular economy—and of society, who should change its consumer choices in

[e]See E. Bulmer's chapter on circular economy in this book for a deeper understanding of this economic model within the EU.

favor of those more environmentally friendly and in line with circularity. That is why a change in the current production model must be made in full collaboration with the main industries and businesses so that such change can have a real impact. In this sense, in the documents analyzing the possibilities of this new economic model, the Commission establishes a direct relationship between economic growth—which will come from the jobs generated by the circular economy—and the positive impact that it will have on the environment (European Commission, 2015). In this respect, circular economy could reduce CO_2 emissions by 48%, create a net economic benefit of €1.8 trillion, and two million additional jobs in the EU by 2030 (Ellen MacArthur Foundation, 2015; European Commission, 2014a). In addition to all these benefits, commitment to this new production model will help the EU to have greater weight as a leader in climate governance and to accelerate its commitments to reduce GHG emissions (Ruiz-Campillo, 2018). The *2020 Strategy* sets three priorities: smart, sustainable and inclusive growth in a Europe that pollutes less, is more energy efficient and is a greater renewable energy producer. What the European Commission was then proposing was to boost the European economy taking into consideration the decisions that had been adopted by Member States only one year before. All of which would reinforce the EU's leadership not only in the environmental arena, but also in the promotion of a more sustainable economy.

As pointed out above, in 2011, after the Copenhagen and Cancun Agreements, the EU Commission adopted a plan to promote a low carbon economy (European Commission, 2011). Three years after, in September 2014, this time ahead of the 21st COP in Paris, the European Commission proposed a new program to boost a circular economy in Europe, reinforced a year later with the document *Closing the loop—An EU action plan for the Circular Economy*. At the end of 2015, the Commission adopted the circular economy package and, since then, different legislative initiatives have been proposed, such as increasing the rate of recycling, packaging waste or promoting the manufacture of products of higher quality that are more durable, repairable and recyclable (European Commission, 2015). If this model was to be fully implemented in the EU, an increase of resource productivity by 15% between 2014 and 2030 under a business-as-usual scenario is forecasted, and in a context of a more circular economy it could double that rate (European Commission, 2015), impacting not only on the access to natural resources, which is expected to satisfy between 10% and 40% of the demand for raw materials of the EU through a better legislation on waste and recycling (European Commission, 2014b), but also on the reduction of GHG emissions through a better waste management—a sector that represents 3% of the EU's emissions (Eurostats, 2018). These are, indeed, the two areas where Member States have made the greatest efforts regarding the implementation of the circular economy. The EU has since 2018 upgraded legally binding targets for recycling, packaging and waste management. Waste directive 2018/851, for instance, incorporates the clear objective of managing waste to contribute to the principles of a circular economy and through it European states have committed to increase to a minimum of 65% the reuse and the recycling of municipal waste by 2035, as well as to set up separate collection for at least paper, metal, plastic and glass by 2021 and for textiles by 2025. Directive 2018/52 on packaging and packaging waste, on its part, establishes specific targets for recycling 70% of all packaging by 2030, and through Directive 2018/850 on the landfill of waste, Member States have agreed to reduce to 10% or less the total amount of municipal waste generated by 2035.

Together with the Energy Union, the promotion of a circular economy is seen as a means to reduce GHG emissions through a better use of raw materials, the promotion of a clean production model in industries and the boost of the waste management sector. In both sectors—energy and economy—the EU not only seeks to reduce emissions, as it has also found that they can be of utmost importance in the improvement of the European economy. At the same time, promoting the use of renewable and a low-carbon economy at the international level can be seen as two ways to continue leading by example in the adoption of sound sustainable and environmentally friendly internal policies.

Conclusions

The climate change crisis has to be seen as an opportunity to make our lives better and not as a burden to our development. The chapter has tried to shed light on how international agreements and European leadership have led the EU to adopt domestic goals and how those goals are slowly transforming areas such as the economy or the energy system in Europe to adapt to more sustainable and cleaner forms of production. With a normative approach behind its actions, the EU sees in this transformation a means to achieve its emission neutrality by 2050 as well as an opportunity to show leadership in the international domain and 'reap the benefits of first mover advantage' (European Commission, 2018a). The EU will have to keep abreast of the exciting times we are living and uphold both the international and European commitments it has made if the EU wants to continue leading environmental negotiations. What it seems clear is that leadership in this area is closely linked to meeting ambitious goals in an area that has to be seen as a means to create a better, more responsible, less polluted, more social and more sustainable economy in Europe.

The EU has proved to be able to adapt to international circumstances in order to keep its leadership. As seen, it responded with stronger diplomatic tools after the COP in Copenhagen—when a second fiddle role began for Europeans—it responded with proposals when Kyoto's top-down hierarchy changed to a United States–China's preferred bottom-up architecture, and although the Paris conference could be seen as a failure of the EU to raise international commitments, the readjustment of its ambition can also be analyzed as an ability to adapt to the international context so as to maintain its leadership in negotiations and reach out to a greater number of parties (Oberthür and Groen, 2017). The leading role currently being played by the EU in environmental negotiations cannot be explained, however, without taking into account the international context in which negotiations take place, the support of Members States, the transformation of the EU's industrial and economic models, the always ambitious initiatives of the European Commission and the good progress being made in the reduction of GHG emissions as well as the promotion of renewable energy. All these ambitious policy proposals are strengthening the EU's leadership to the point that some believe that they may have played a role also in raising other countries' emissions reduction ambitions (Schunz, 2011). It is clear that the transition to a carbon neutral Europe will be painful in many senses; however, the EU knows that there is no alternative to this and has to step up its efforts to continue being a leader in the international community and an example and supporter to other countries with fewer resources.

The chapter has examined how the EU has reinforced its international leadership from the inside, adopting ambitious targets internally, but also being strategic in their adoption. In some cases those targets have been the response to international agreements, but in others they have been intended to show European environmental compromise in international climate negotiations. It is clear that the EU is stronger when acting united, and a pattern seen before almost any main climate conference is the internal adoption of ambitious initiatives, policies or communications from the European institutions as a means of being proactive and leading by example in international negotiations.

With the adoption of the 2020 and 2030 targets in 2007 and 2014, respectively, the EU is regulating the reduction of GHG in all economic areas and thus strengthening its leadership in environmental negotiations by showing other countries its ability to adopt ambitious targets. The results of the actions adopted by Member States are already visible: in 2016, the EU had already reduced its emissions by 22.4% below 1990 levels, partially thanks to structural changes in the economy, reduced economic activity due to the economic crisis in 2008 and improved energy efficiency (EEA, 2018). Although it seems that emissions will be further reduced by 2020, the truth is that since 2014 progress in reducing emissions has slowed down. Hence, reaching European 2030 goals may be more difficult than expected. As for the goals in renewable energy sources, in 2016 the share reached 17% of the gross final consumption in the EU, but contrary to what happens with the reduction of GHG, the increase of renewable energy has been slower than expected (EEA, 2018). Explanations for this go from opposition to the promotion of biofuels or the acute economic crisis that severely impacted the development of renewable energy in European countries, especially in Spain, Portugal or Greece (Fisher and Geden, 2013).

The promotion of the Energy Union and of the circular economy on their part is the result of the EU's strive to reduce its emissions in an area in which it has found a natural leadership. The adoption of binding targets is fundamental to undergo this transformation and despite the difficulties in the adoption of common targets among 28 Member States, the EU has been remarkably ambitious. As has been explained, the goals to reduce GHG and increase renewable energies have entailed the adoption of an emissions trading market, the promotion of greater recycling targets, of greater efficiency, of renewable energy, of investments in industrial processes and the mobilization of cities and towns to achieve all this. This transformation in return will back European leadership in international climate negotiations with facts, but above all will improve the quality of the lives of European citizens with less polluted countries.

***

While this book was in the final stage of its editing, the new European Commission, led by Ursula von der Leyen, launched the European Green Deal (December 2019) at the time the COP25 was being held in Madrid. The Green Deal reinforces European leadership at the international level by setting a credible example to the rest of the world. But it also deepens on all the strategies and policies to reduce GHG this book describes along its chapters. What the Green Deal clearly shows is that there is no return on the fight against climate change in the EU and on the internal transformation that it will require to face it. The new proposal increases the EU's greenhouse gas emissions reduction target to at least 50% and toward 55% compared with 1990 levels which means that further transformations will be needed in the economy, the industry and the energy systems. Probably, the greatest achievement of this new proposal is the mainstreaming it makes of climate change in all EU policies, going from economy and energy to transport, agriculture and farming, research or education amongst others. Because if climate change is to be faced efficiently it can only be done from a holistic and multisectoral approach coming from above.

References

Afionis, S. and Stringer, L. (2012) European Union leadership in biofuels regulation: Europe as a normative power? *Journal of Cleaner Production*, 32, pp. 114–123.

Barbé, E., Herranz-Surrallés, A. and Natorksi, M. (2015) Contending metaphors of the European Union as a global actor. Norms and power in the European discourse on multilateralism. *Journal of Language and Politics*, 14(1), pp. 18–40.

Behrens, A. (2016) Time to connect the dots: What is the link between climate change policy and the circular economy? *CEPS Policy Brief Nº 337*.

Chasek, P., Downie, D. and Welsh brown, J. (2014) *Global Environmental Politics*. 6th edition. Westview Press, Colorado.

Commission of the European Communities (1991) *A Community Strategy to limit Carbon Dioxide Emissions and to Improve Energy Efficiency*. Communication from the Commission to the Council, 14 October 1991.

Commission of the European Communities (2007) *Limiting Global Climate Change to 2 Degrees Celsius. The Way Ahead for 2020 and Beyond. Communication from the Commission*, 10 January 2007.

Delbosc, A. and Perthuis, C. (2009) *Carbon Markets: The Simple Facts.* Caring for Climate Series, UN Global Compact Office, Paris, France.

Deloitte (2016) *Circular Economy Potential for Climate Change Mitigation.* Deloitte Conseil, London, UK.

EEA (2016) *Renewable Energy in Europe 2016. Recent Growth and Knock-on Effects*, EEA Report n° 4/2016.

EEA (2018) *Annual Indicator Report Series (AIRS).*

Ellen MacArthur Foundation (2015) *Growth Within: A Circular Economy Vision for a Competitive Europe.*

Ellerman, A., Marcantonini, C. and Zaklan, A. (Winter 2016) The European Union Emissions Trading System: Ten Years and Counting. *Review of Environmental Economics and Policy*, 10(1), 89–107.

Eurobarometer (2008) *35 Years of Eurobarometer. European Integration as Seen by Public Opinion in the Member States of the European Union 1973–2008*, European Commission.

Eurobarometer (2017) *Attitude of European Citizens towards the Environment.* Special Eurobarometer, European Commission.

European Commission (1997) *Energy for the Future: Renewable Sources of Energy. White Paper for a Community Strategy and Action Plan. Communication from the Commission*, 26 November 1997, available at: http://europa.eu/documents/comm/white_papers/pdf/com97_599_en.pdf

European Commission (2010) *Europe 2020. A European Strategy for Smart, Sustainable and Inclusive Growth. Communication from the Commission*, 3 March 210.

European Commission (2011) *A Roadmap for Moving to a Competitive Low Carbon Economy in 2050. Communication from the Commission*, 8 March 2011.

European Commission (2014a) *Study on Modeling of the Economic and Environmental Impacts on Raw Material Consumption.*

European Commission (2014b) *Towards a Circular Economy: A Zero Waste Programme for Europe. Communication from the Commission*, 25 September 2014.

European Commission (2015) *Closing the Loop — An EU Action Plan for the Circular Economy*, 2 December 2015.

European Commission (2016) *Clean Energy for all Europeans. Communication from the Commission*, 30 November 2016.

European Commission (2018a) *A Clean Planet for all. A European Strategic Long-Term Vision for a Prosperous, Modern, Competitive and Climate Neutral Economy. Communication from the Commission*, 28 November 2018.

European Commission (2018b) *A Stronger Global Actor: A More Efficient Decision-Making for EU Common Foreign and Security Policy. Communication from the*

Commission to the European Council, the European Parliament and the Council, Brussels, 12 September 2018.

European Commission (2019) *EU ETS 2005–2012. Phases 1 and 2*, available at: https://ec.europa.eu/clima/policies/ets/pre2013_en, Consulted on April 2019.

European Council (1990) *Presidency Conclusions. The Environmental Imperative* (Annex II), Dublin, 25 and 26 June 1990.

European Council (2007) *Presidency Conclusions 8/9 March 2007*, Brussels, 2 May 2007.

European Union (2013) *General Union Environment Action Programme to 2020 'Living Well, within the Limits of our Planet'*, Decision N° 1386/2013/EU of the European Parliament and of the Council of 20 November 2013.

European Union (2016) *Supporting Study for the Evaluation of Decision N° 406/4009/EC (Effort Sharing Decision). Final Report*, Luxembourg.

Eurostats (2018) *Greenhouse Gas Emission Statistics – Emission Inventories. Statistics Explained*. Consulted in April 2019.

Falkner, R. (2007) The political economy of 'normative power' Europe: EU environmental leadership in international biotechnology regulation. *Journal of European Public Policy*, 14, 4.

Fisher, S. and Geden, O. (2013) *Updating the EU's Energy and Climate Policy. New Targets for the Post-2020 Period*. Friedrich Ebert Stiftung, Berlin, Germany.

Hettne, B. and Söderbaum, F. (2005) Civilian power or soft imperialism? The EU as a global actor and the role of interregionalism. *European Foreign Affairs Review*, 10, 535–552.

Howorth, J. (2010) The EU as a global actor: Grand strategy for a global grand bargain? *Journal of Common Market Studies*, 48(3), 455–474.

Joint Research Centre (JRC) (2016) *Covenant of Mayors: Greenhouse Gas Emissions Achievements and Projections. Science for Policy Report*, Italy, European Union, available at: https://publications.jrc.ec.europa.eu/repository/bitstream/JRC103316/jrc103316_com%20achievements%20and%20projections_online.pdf

Kelemen, D. (3 April 2010) Globalizing European Union environmental policy. *Journal of European Public Policy*, 17, 335–349.

Kulovesi, K. (2012) Climate change in EU external relations: Please follow my example (or I might force you to), in Morgera, E. (ed.), *The External Environmental Policy of the European Union. EU and International Law Perspectives*. Cambridge University Press, Cambridge, UK.

Manners, I. (2002) Normative power Europe: A contradiction in terms? *Journal of Common Market Studies*, 40(2), 235–258.

Manners, I. (2011) The European Union's normative power: Critical perspectives and perspectives on the critical, in Whitman, R. (ed.), *Normative Power Europe. Empirical and Theoretical Perspectives*. Palgrave Macmillan, Basingstoke, UK.

Mayer, H. (2008) The long legacy of Dorian Gray: Why the European Union needs to redefine its role in global affairs. *European Integration*, 30(1), 7–25.

McDonough, W. and Braungart, M. (2002) *Cradle to Cradle: Remaking the Way we Make Things*. North Point Press, New York.

Milios, L. (2018) Advancing to a circular economy: Three essential ingredients for a comprehensive policy mix. *Sustainability Science*, 13, 861–878.

New Scientist (2007) *Europe's '2020 vision' to Lead Climate Change Battle*, 9 March 2007.

Oberthür, S. and Groen, L. (2017) The European Union and the Paris agreement: Leader, mediator, or bystander? *WIREs Clim Change*, 8, e445.

Oberthür, S. and Pallemaerts, M. (2010) *The New Climate Policies of the European Union. Internal Legislation and Climate Diplomacy*. Brussels University Press, Brussels, Belgium.

Oberthür, S. and Roche Kelly, C. (2008) EU leadership in international climate policy: Achievements and challenges. *The International Spectator*, 43(3), 35–50.

OJEU (2018a) Directive (EU) 2018/2001 of the European Parliament and of the Council of 11 December 2018 on the promotion of the use of energy from renewable sources.

OJEU (2018b) Directive (EU) 2018/2002 of the European Parliament and of the Council of 11 December 2018 amending Directive 2012/27/EU on energy efficiency.

OJEU (2018c) Regulation (EU) 2018/1999 of the European Parliament and of the Council of 11 December 2018 on the Governance of the European Union and Climate Action.

Ruiz-Campillo, X. (2017) Liderazgo y diplomacia de la Unión Europea en las negociaciones climáticas, in Martínez Capdevila C. and Martínez Pérez E. (eds.), *Retos para la acción exterior de la Unión Europea*. Tirant Lo Blanch, Valencia, Spain.

Ruiz-Campillo, X. (2018) La apuesta de la Unión Europea por el desarrollo sostenible: de la economía circular al Acuerdo de París, in Juste, J. and Bou, V. (ed.), *Desarrollo Sostenible y Derecho Internacional*. Tirant Lo Blanch: Valencia.

Schunz, S. (1 January 2011) Beyond leadership by example: towards a flexible European Union foreign policy, *Working Paper FG 8*.

Sitra et al. (2018) *The Circular Economy. A Powerful Force for Climate Action. Transformative Innovation for Prosperous and Low-Carbon Industry*. Material Economics, Stockholm, Sweden.

Thieffry, P. (2016) Environmental protection and European Union enrgy policy: Energy transition after the Paris Agreement. *ERA Forum*, 17, 449–465.

Torney, D. (October 2013) *European Climate Diplomacy. Building Capacity for External Action*, FIIA Briefing Paper 141.

Tuerk, A. and Zelljadt, E. (2016) *The Global Rise of Emissions Trading.* Climate Policy Info Hub, 11 April 2016, available at: http://climatepolicyinfohub.eu/global-rise-emissions-trading

Vogler, J. and Stephan, H. (2007) The European Union in global environmental governance: Leadership in the making? *International Environmental Agreements,* 7, 389–413.

Whitman, R. (2011) *Normative Power Europe. Empirical and Theoretical Perspectives.* Palgrave Macmillan, Basingstoke, UK.

Zito, A. (2005) The European Union as an Environmental Leader in a Global Environment. *Globalizations,* 2(3), 363–375.

CHAPTER 2

The European Union and Its Role in Climate Change Negotiations at the UNFCCC: From a Loss to a Recovery of Leadership with Costs 2009–2018

María del Pilar Bueno

The Community and its Member States have a special responsibility to encourage and participate in international action to combat global environmental problems. Their capacity to provide leadership in this field is enormous.

–European Council (1990)

Introduction

A key feature of this book is to understand, explain and provide details on the European Union (EU) leadership in climate change policy at the international level, in particular, at the United Nations Framework Convention on Climate Change (UNFCCC) since its adoption at Rio Summit in 1992. This international leadership has been built on the basis of regional solidarity in the distribution of mitigation efforts within the EU. This can be understood as a specific interpretation of the principle of common but differentiated responsibilities (CBDR) (Bueno, 2014, 2017).

This distribution of efforts as a result of the commitment allocated in the Kyoto Protocol (KP) has been explained by Delbeke and Vis (2015) as cost-effective and fair.

However, we wonder whether this leadership has been originally a policy objective of European climate change diplomacy or it was a result that became at a second-round a policy option with the blessing of civil society organisations and, more recently in time, businessman support. A new vision that, after the KP's first failure with US withdraw, was based on climate science (IPCC, 2007),[a] the economics of climate change and the Stern Review (2007) and the association of the climate and energy package. At the same time, this package allowed Europe to dream and aspire to a future of energy autonomy grounded on energy transition and oiled by low-carbon economy flows.

Readers may wonder if it is important or not to know what was first, the chicken or the egg. The phrase included in the beginning of the chapter is intentional to provoke the question on the *when*, even if this contribution is more focused on the *how*.

What it is worthy to understand is how the EU benefited from a top-down approach adopted at the international level during first phase of UNFCCC experience, since it allowed the Union to distribute the commitments according to this fairness and cost-effectiveness approach.

All the ingredients of the recipe already mentioned, as well as many others, helped the EU to build a successful community climate policy that took it to the podium of compliance of the 8% of emission reduction committed with the KP. For achieving so, the EU fostered the most successful Emissions Trading System (EU ETS) even when market mechanisms under Kyoto had been a US initiative to which EU had shown initially reluctant. As Bodansky (2010, 2011) expressed, EU pushed for strong emission reduction targets—since Noorwijk Conference in 1989—founded on domestic measures and United States with the Umbrella Group pushed for the use of market-based mechanisms. However, this first phase in climate change regime negotiations came to its end with Copenhagen failure allowing to a new era of adjustment where the United States and the Chinese attitudes towards the international regime changed forcing the EU to compete for the leadership.

Although the Copenhagen failure required a new faith in the international regime and the EU was again a critical actor to achieve so, the progressive shared leadership at the UNFCCC arena demanded the EU

[a]The Third Report of the Intergovernmental Panel of Climate Change (AR 4 IPCC, 2007) was published on time for the 13 Conference of the Parties of the UNFCCC celebrated in Bali. The Bali Mandate adopted in 2007 mandated the negotiations for a new legal document to be adopted in COP 15 in Copenhagen, 2009.

to preserve internally its old recipe. That means that the EU retained the energy–climate package, invested more in technology and energy transition, improved the ETS even with difficulties, continued to push for carbon pricing and for the decoupling of emissions from economic growth.

From Copenhagen to Durban, the triangular leadership formed by United States, China and the EU agreed on a mandate that shaped the conditions for the new agreement. Nevertheless, different models or architectures clashed as explained by Bodansky (2011). On the one hand, a *top-down approach* similar to the KP where the regime established the commitments for Annex I Parties while providing some space for countries to define how to implement those. On the other hand, a *bottom-up approach* where each Party can define flexibly how to contribute to the regime. However, the urgency of effective and collective climate action based on science shows that correct incentives should accompany the second model in order to make it possible and compatible with the global challenge we face. What we agree with the author aforementioned is that Copenhagen–Cancun meetings provided to the international regime an opportunity to unblock a stagnant process (Bodansky, 2011).

Nevertheless, while the United States and China built an alliance based on a bottom-up logic in Copenhagen, the EU did not abandon its aspiration based on a top-down model that allowed it to sustain its own domestic process of distribution of efforts. These two visions can be traced back to the Paris Agreement (PA) and even during the first implementation period. At the same time, these two models or architectures are based on different foundations, since EU top-down understanding of climate regime is grounded on collective action traditional dilemma where states can only be compelled to act when others do the same avoiding the free riding problem. United States and others have argued that US subnational climate policy have progressed even when federal policy did not.[b] Besides, the United States could not accept and international agreement with binding commitments without including China. In addition, some of the most

[b]The main idea here is to contract collective action dilemmas by saying not necessary a legally binding agreement is necessary with specific commitments is necessary to compel action and solve climate change since there are many public and private actors acting against climate change that are not compelled to do it but there are other incentives to pay attention.

important emitters of developing world, such as China, were not disposed to accept binding mitigation commitments that could reinterpret the differentiation under the Convention. For those reasons, a bottom-up and flexible approach could be more compatible with apparently irreconcilable differences between United States and Chinese interests.

As a result, we propose to analyse the changes experienced in EU leadership in three moments from 2009 to 2018 by contrasting what we understand as important features of climate leadership at the international arena and using the available and substantial-related literature.[c] These features include *vision* of the actor about the international architecture (Parker et al., 2012); *internal coherence*, wondering whether the positions and proposals are based on the national/regional experiences (Oberthür, 2011; Groen et al., 2012; Bäckstrand and Elgström, 2013); *degree of achievement of the objectives and proposals* formulated (Parker et al., 2012; Parker et al., 2017) and the *flexibility to interpret and adapt to unstable environments* (Oberthür, 2011; Afionis, 2011; Bäckstrand and Elgström, 2013; Costa, 2016).

In the following pages, we develop an analysis of the EU leadership, first, applied to the Copenhagen Accord, as well as the Cancun Summit and the Durban Platform, in the context of the second phase of progressively shared leadership with the United States and China. Second, we analyse the Paris Conference and the results obtained, in particular, the PA and Decision 1/CP.21. Finally, we examine the first period of implementation of the PA from May 2016, when the agenda of the Ad Hoc Working Group of the Paris Agreement (APA) was negotiated and adopted, to the COP24 celebrated recently in Poland. At the same time, the conclusion tends to provide some ideas about current and future challenges to the EU leadership at the UNFCCC arena.

[c]There is a tradition in this discussion of EU actorness and leadership in climate change negotiations. Some authors that honor this tradition are Vogler, 1999; Yamin, 2000; Gupta and Grubb, 2000; Gupta and Ringius, 2001; Andresen and Agrawala, 2002; Zito, 2005; Vogler and Bretherton, 2006; Groenleer and Van Schaik, 2007; Schreurs and Tiberghien, 2007; Compston and Bailey, 2008; Damro and MacKenzie, 2008; Oberthür, 2009; Wurzel and Connelly, 2010; Afionis, 2011; Groen et al., 2012; Parker et al., 2012; Van Schaik and Schunz, 2012; Bäckstrand and Elgström, 2013; Costa, 2016; Parker and Karlsson, 2017; Parker et al., 2017.

2.1 The Sequence Copenhagen–Cancun–Durban and the Beginning of the Shared Leadership

Decision 2/CP.15 took note of the Copenhagen Accord of 18 December 2009.[d] The Accord recognised that according to the science, and in particular IPCC AR4 (2007), the increase in global temperature should be below 2°C above the pre-industrial levels, according to EU position. However, it did not develop specific commitments that could be distributed at the international as well as the European level following a top-down approach. So it was more aspirational in terms of emission reductions and responsibilities and imprecise according to European expectations.

It was recognised that a peaking in global emissions should take place as soon as possible and it will take more time in developing countries. It is important to highlight that this is a language acceptable for developing countries even when many of the developed countries had already achieved the peaking at that time, so the peaking language required more action of developing countries with increasing emissions.

Moreover, the Accord did not break the so-called firewall between developed and developing country Parties enriched by differentiation since it recognised the principle of CBDR and respective capabilities (CBDR-RC) as well as equity. However, the EU had emphasised that it respected the CBDR-RC principle and it was the United States and the Umbrella Group the most interested in its decline.

In terms of mitigation efforts, the Copenhagen Accord disposed that the Annex I Parties should submit before 31 January 2010 to the Secretariat of the Convention, quantified economy wide emission targets for 2020 for its compilation. For Parties with mitigation responsibilities under the KP, those targets should further strengthen in comparison with previous efforts. This was a clear antecedent of the progression principle under the PA. Furthermore, these actions had to be subject to international monitoring, reporting and verification (MRV) according to guidelines to be

[d]The report of the Conference of the Parties on its 15th session, held in Copenhagen from 7 to 19 December 2009, as well as the Addendum and decisions adopted by the Conference of the Parties can be consulted in https://unfccc.int/resource/docs/2009/cop15/eng/11a01.pdf

determined in future meetings. For developing countries, the Accord also established that they had to submit in the same date, mitigation actions, subject to domestic MRV. This last provision was a condition imposed by China and strongly opposed by EU since transparency requirements were critical for the Europeans. These actions were voluntary for least developed countries (LDC) and small developing island states (SIDS). The Accord also allowed developing countries to prepare and submit nationally appropriate mitigation actions (NAMA) seeking of support in terms of finance, technology and capacity building.

The European disappointment with the mitigation outcome originates in the objectives and official position that the developed countries' commitment needed to amount to a cut in collective emissions in the order of 30% below 1990 levels by 2020. As well as more economically advanced developing countries had to pledge ambitious, quantified mitigation actions showing a substantial deviation, in the order of 15%–30% below, the currently predicted growth rate in their collective emissions by 2020 (European Commission, 2009).

Coming back to the transparency provisions, the Accord created a new international consultation and analysis (ICA) process for the Biennial Update Reports (BUR) of the developing countries compatible with the international assessment and review (IAR) of the Biennial Reports (BR) of the developed countries. This was critical for the EU since one of the most important parts of the architecture sought by the Europeans was focused on a strong transparency system. This system, while recognizing the different starting points, had to equate the accountability actions to a common system based on MRV of mitigation actions. However, this improvement in transparency concurring with the EU standards was not good enough according to the flexible pledges announced before the meeting by both developed and developing countries. These pledges, submitted again by 31 January, were supposed to be part of the aforementioned compilation.

In terms of climate finance, the document compromised funding for mitigation and adaptation actions of developing countries, including a fast track of 30 billion dollars between 2010 and 2012 and a goal by 2020 of mobilizing 100 billion per year. EU has been continuously disposed to accomplish its commitments to provide support to developing countries, but under certain circumstances, including public–private sources, as well as bilateral and multilateral funding and alternative sources. The EU also emphasised the intention of focusing the support on low-income countries, an inexistent category in the UNFCCC and not acceptable for developing countries. So that, SIDS and LDC were the countries that had to be prioritised according to the European position. Nevertheless, the Accord disposed of the finance, as well as technology and capacity building, had to be able for developing countries without distinctions.

During COP15, the official position stated that both mitigation and financial commitments should be captured in the agreement in order to distribute the contributions in a fair share. European Commission compromised 22–50 billion Euros per year by 2020 to achieve the global agreement (European Commission, 2009).

The adaptation was also part of the Accord but being very vague, just underlined the challenge of adaptation and the importance of developed countries to provide support to developing countries for their adaptation actions. Adaptation at the regional level was very important for the EU but was not a priority to negotiate at the international level. However, the EU supported the establishment of an adaptation framework (European Commission, 2009).

Finally, the Copenhagen Accord also established a Mechanism including the Programme of Reducing Emissions from Deforestation and Degradation (REDD plus), as well as, a technology Mechanism for technology development and transfer both supported by the EU (European Commission, 2009).

The Accord was rejected by Sudan, Venezuela and Bolivia breaking the consensus rule, milestone of UNFCCC negotiations. This was the end (it was suspended) of a COP presided by a European country where, for the first time, developing countries showed that the lack of transparency and participation in decision-making processes was not admissible at the multilateral level.

The main goal of the EU was to achieve a comprehensive, ambitious, fair, science-based and legally binding global treaty to succeed the KP (European Commission, 2009). Consequently, the failure in Copenhagen was more tragic for the EU in comparison with other countries, and in particular the United States, since it saw the decline of its leadership at least, as it used to be before. In addition, it was the host of the Conference, as well as it was the most compromised with the multilateral process in absence of the United States of the KP.

Since the bottom-up approach triumphed, and President Obama as well as the BASIC Presidents were the star figures of the COP,[e] it was necessary to reconsider the strategies and how European interests could be interwoven with United States and Chinese priorities and divergences. As many authors highlighted during the final stages of the negotiations, the EU Presidents were marginalised of decision-making process starring by United States and BASIC Presidents (Curtin, 2010; Van Schaik and Schunz, 2012; Groen et al., 2012; Parker et al., 2012).

Copenhagen was not an ending point in terms of a global treaty, Cancun negotiations at COP16 held in Mexico were critical in order to revitalise the multilateral climate process as well as the European leadership (Oberthür, 2011). In that way, Groen, Niemann and Oberthür affirmed: "the EU seems to have played a more influential role at the Cancun negotiations than at Copenhagen, being more involved in the decision-making process and having a firmer grip on the outcomes," but for doing so, it had less ambitious and more pragmatic objectives and approach (Groen et al., 2012: 2). Not only that, as affirmed by the authors, the EU benefited from not being the hosts of the meetings, as well as, of the generalised need of agreement, in particular of United States and BASIC countries, that showed themselves more cooperative than one year before.

The foremost success of Cancun was to collect main points of the Copenhagen Accord and turn them into decisions, including the establishment of the Green Climate Fund and the Cancun Adaptation Framework (CAF), strongly enriched in comparison with the COP15 outcome. The CAF instituted the Adaptation Committee to promote the implementation of enhanced action on adaptation in a coherent manner with the Convention, as claimed by the developing countries for many years. In addition, other specifications were provided in terms of ICA process, the technology mechanism and REDD plus.

---

[e]The summary note of the Parliament's delegation to the COP15, expressed that "the European Union was not part of the meeting between the US, China, India, Brazil and South Africa, in which the accord, as negotiated by the 27 Heads of State and Government, was further weakened. The European Union should learn from Copenhagen and should ensure that it will be able to play a leading role in Mexico." The summary note of the EP members at COP15 is contained in http://www.europarl.europa.eu/document/activities/cont/201001/20100122ATT67849/20100122ATT67849EN.pdf

All of these points were supported by the EU, as stated by Connie Hedegaard, EU Commissioner for Climate Action and Joke Schauvliege, Flemish Minister for the Environment representing the Belgian Presidency during the last negotiation session before COP16 celebrated in Tianjin, China: "There is a strong will that the EU will continue its leadership in the fight against climate change and will speak with one voice in Cancun. In Cancun, we want to see a balanced set of decisions (...) In particular, we want to see concrete decisions on Forestry, Adaptation, Measurement, Reporting, Verification (MRV) and on Technological Development. We can assure the world that the EU will deliver on the fast start funding. The Parties must build on the Copenhagen Accord."[f, g]

The final stop of the sequence is Durban Conference held in South Africa in 2011. COP17 was the first meeting to be developed in a BASIC country after the beginning of its alliance and Copenhagen failure. Consequently, the cooperative spirit of the group that emerged in Cancun meeting continued to be present during Durban Conference.

In terms of the EU priorities, it is important to identify the main points included in the EU statement during the opening ceremony of the High Level Segment where Connie Hedegaard and Minister Korolec from Poland reaffirmed the need of progress in terms of the gap between the 2°C target and the pledges to achieve so. Moreover, the representatives emphasised the will to advance to a second period of commitments of the KP and a roadmap for a new comprehensive, legally binding and global agreement, as well as the importance of finance and transparency.[h] The official EU position also comprised that the elements to include in the new Protocol or agreement should be the ones agreed in Bali and Cancun. That is the main reason why the representatives mentioned mitigation, finance and transparency as their priorities, even when EU recognised adaptation and

[f]The Press release can be consulted at https://www.eea.europa.eu/highlights/cop16-climate-change-talks-start

[g]These priorities appear also in the interventions made by the representatives of the EU in the opening ceremony of the High Level Segment of the COP16 in Cancun, available in https://unfccc.int/files/meetings/cop_16/statements/application/pdf/101209_cop16_hls_eu.pdf

[h]The intervention can be consulted in https://unfccc.int/files/meetings/durban_nov_2011/statements/application/pdf/111206_cop17_hls_european_union.pdf

technology as integral part of the Bali Road Map. Furthermore, the EU reaffirmed the importance of market-based mechanisms as well as the recognition that even when mitigation commitments should respect CBDR-RC principle, it had to reflect evolving global political and economic realities (European Commission, 2011). This idea of evolving interpretation of the principles of the Convention and, in particular of the cornerstone of the differentiation system under the UNFCCC, was one of the most important battles in the climate multilateral arena from 1992.

Another key point of the EU position was the development of accounting rules for forest management and emission from land sector during the second period of commitments of the KP. Moreover, EU pushed for the introduction of aviation and maritime transport sectors not covered by the KP.

The Durban Platform for Enhanced Action was established by the decisions adopted in COP17,[i] and according to the European expectations, it provided the mandate to achieve no later than 2015 a protocol, another legal instrument or an agreed outcome with legal force under the Convention applicable to all Parties. To carry out the work, the Platform instituted the Ad Hoc Working Group on the Durban Platform for Enhanced Action (ADP). This instrument was expected to come into effect from 2020.

The elements included in the mandate were mitigation, adaptation, finance, technology development and transfer, transparency of action and support and capacity-building. All of them were acceptable for the EU position.

The Platform also progressed in terms of the second period of commitments of the KP that Doha Amendment was adopted in Doha in 2012.

As recognised in the official position of the EU, additional technical work in terms of the institutions took place, including the Adaptation Committee, the Technology Mechanism, the Standing Committee of Finance and the Green Climate Fund. Initial guidelines of National Adaptation Plans included in the CAF were also developed as well as some progress was made in terms of REDD plus.

---

[i]The Report of the Seventeenth Session of the Conference of the Parties as well as all the decision adopted during COP17 are available in https://unfccc.int/resource/docs/2011/cop17/eng/09a01.pdf

In mitigation, the expectations needed to be moderated, and same case with land sector (LULUCF) rules as well as aviation and maritime emissions. However, transparency measures progressed, including the adoption of modalities and procedures of IAR and ICA processes.

Summarising this section and trying to tend bridges with diverse approaches to European leadership in climate change negotiations, different authors (Groen et al., 2012; Bäckstrand and Elgström, 2013) agree on the idea that the EU position in Copenhagen was not flexible enough and too aggressive in terms of emission reductions even though it was based on its own efforts and achievements at the community level. The internal coherence was good since it was guided by the adoption of the 20/20/20 package in the EU in 2007. Nevertheless, this confrontational strategy was not functional to a host of an international conference seeking for an agreement. As a result, the degree of achievement of the proposals was very low since the most important goal was to reach an agreement.

Furthermore, the position was a result of a misguided interpretation of the international context, in particular, the changes occurred in the US administration and the BASIC group alliance. At the same time, the diplomats and ministers did not have the necessary margins and flexibility to move from the original positions.

Finally, the vision that guided the architecture for so many years began to see its decline at the hands of the bottom-up system that was shown as the hope to get out of the stalemate of the process.

Based on what Delbeke and Vis (2015) understand as a "learning by doing process," EU climate diplomacy could accommodate to these new conditions and assumed a more pragmatic approach in Cancun and Durban. This helped the EU to accomplish its priorities and revitalise and recovery the leadership (Verolme, 2012; Wu, 2012). As affirmed by Bäckstrand and Elgström (2013), the EU shifted from a normative and ideational leadership to a more pragmatic and structural one, looking for bridges among Parties. One example in terms of those bridges was to abandon the more aggressive diplomacy with developing countries followed in Copenhagen to look for coalition formation strategies through the Cartagena Dialogue for Progressive Action, composed by 30 developed and developing countries (Oberthür, 2011; Bäckstrand and Elgström, 2013).

2.2 The EU Leadership Towards the PA

The journey towards the adoption of the PA in December 2015, according to the Durban Platform and the ADP ups and downs, had many pitfalls and difficulties for all the Parties. From 2011 to 2015, the United States and China reaffirmed their alliance with very specific details that changed the results of the COP from Durban to Paris including the pillars of the future agreement and the regime. One of the most important examples is the Joint Announcement on Climate Change as a result of the meeting between Presidents Barack Obama and Xi Jinping in Beijing in November 2014. In the Announcement, the Presidents reaffirmed their commitment to reach an ambitious agreement according to the Durban Platform that reflected the principle of CBDR and respective capabilities, in light of different national circumstances. The amended principle of CBDR was some days later adopted by Decision 1/CP.20 in Lima safeguarding the way in which differentiation could be acceptable for the big players in the architecture of Paris.[j] Moreover, it was adopted in the PA with the same format.

There are many other examples but the focus here is the European leadership, so that, it is critical to understand to what extent the alliance United States–China forced the EU to adjust its own positions without losing its main interests.

In Paris, one more time, the EU was host of an international conference, but this one promised to be memorable and indeed it was. The French Presidency looked like a hostess demonstrating its abilities to reach the agreement that seemed impossible but necessary (Dimitrov, 2016). We have analysed the Paris architecture before (Bueno, 2017), by indicating that the bottom-up approach initiated in Copenhagen was nourished by the concept of nationally determined contributions that saw the light in Warsaw in 2013 (COP19). It conquered its first implementation in Lima (COP20) with the idea of Intended Nationally Determined Contributions (INDC) and finally became the main vehicle of climate action in Paris through the NDC (COP21). Not only this, the NDC included all the elements of the PA according to the article 3 even with the opposition of all developed countries.

---

[j] The Joint Announcement is available in https://obamawhitehouse.archives.gov/the-press-office/2015/09/25/us-china-joint-presidential-statement-climate-change, accessed in 14 January 2019.

Focusing on the EU positions towards the negotiations in Paris, it is significant to highlight that the Europeans sought the Summit to be an example of its traditional commitment with multilateralism and the United Nations System. According to their expectations, the 2015 agreement should be ambitious and inclusive (include all Parties and to be according to the climate science), fair (in relation to the evolving principle of CBDR-RC) and robust (strong legal basis with clear compromises). Even when its top-down vision did not have a chance according to the perspective of other actors, the EU supported that a third way was possible as a third model of international climate policy meaning a hybrid between bottom-up and top-down approaches.

The European diplomacy understood the NDC information and the MRV system had to be unified and a common set of rules should be applicable to everyone. So that, the transparency system, as well as the five-year cycles could provide, not only clarity and ambition, but also the top-down perspective that they interpreted was an open door in Warsaw and Lima (Lefevere et al., 2015; European Council, 2015).

The fairness or not of the agreement should be understood in light of each actor's position. If the idea was to delete the firewall between developed and developing countries, the PA maintains and implements the differentiation, but it also makes universal the mitigation action through the NDC as all developed countries required. Furthermore, dynamic cycles where achieved with the Global Stocktake(GST) provisions as the EU proposed.

The EU vision of the transparency system was focused on mitigation targets and national inventories, but it exercised flexibility in terms of including other elements of the Durban Platform. In relation to those elements, it agreed on reinforcing adaptation action and support as well-strengthening monitoring and reporting systems on adaptation. However, it proposed to do so by using the national communications and not the NDC as projected by many developing countries. It should be noted that all developing countries included an adaptation component in their INDC in 2015, even when there were no requirements or instructions in Lima decisions about adaptation in the INDC. Further to the national communications and the NDC, an adaptation communication was created in Paris that could be submitted as a component or in conjunction of those and other documents as referred to in article 7.11 of the PA.

Considering the transparency results and the EU proposal, a common enhanced transparency framework for action and support was adopted including all countries. It also provided flexibilities in consideration of the different starting points of the Parties.

In terms of the mitigation targets, the EU supported the idea of an *ex ante* review of this component in the NDC. Nevertheless, it didn't work since it was one of the most important redlines of the Chinese delegation.

With respect to the long-term goal on mitigation, the European Council (2015) assumed that the Agreement needed to provide a clearly defined pathway to achieving the objective of increasing the global average temperature to 2°C above the pre-industrial levels in order to prevent the worst impacts of climate change. In doing so, emissions had to peak by 2020 at the latest, be reduced by at least by 50% by 2050 compared to 1990 and be near zero or below by 2100. According to the European perspective, this understanding of the distribution of the global efforts was consistent with EU objective of reducing emissions by 80%–95% by 2050 compared with 1990 by developed countries as a group.

The United States and the LMDC countries, in particular, China, were very reluctant to accept mitigation-binding commitments since this was the basis of the agreement between the two aforementioned countries. For that reason, the PA finally included the temperature goal of 2°C that everyone could accept but also comprised the 1.5°C attempt in order to have the small island states on board even with the strong opposition of the Arab countries. It also encompassed the peaking language agreed in Cancun as well as the need to achieve a balance between anthropogenic emissions by sources and removals by sinks of greenhouse gases in the second half of the century. So that, as projected, no specific distribution of efforts was adopted but the leadership of developed countries in mitigation actions and no backsliding was reaffirmed even with the very heavy discussion of the "shall" language that the United States opposed in the last minutes of the Conference.[k]

[k]After the distribution of the last version of the text and the clear signal form the French Presidency that it was a take it or leave it proposal, which was accepted in the last informal session, the formal ceremony was delayed for 90 minutes since the US lawyers said they had not realised before that the last version included language in which developed countries shall continue to take the lead, something that goes beyond the Convention that use a should language for the same idea. After a lot of discussions, the Secretariat of the Convention referred to editorial mistakes and last version changed according to the US needs.

In terms of climate finance, the EU maintained the position of focusing the financial flows in low-emission climate-resilient investments. This includes their commitment to continue financing the mitigation and adaptation activities of developing countries. Nevertheless, it was a redline to incorporate in the agreement a new financial goal.[l] Furthermore, private and public sources, as well as bilateral, multilateral and other innovative sources had to be incorporated even when public finance assumed more relevance as claimed by China, the African Group and many other developing countries (European Council, 2015). In addition, the EU supported in the draft texts the expression: "Parties in the position to do so," applied to developing countries that in light of their evolving capabilities could also provide finance. This language was a reflection of the opposition to maintain the so-called firewall in terms of the donors. Anyway, the expression was a redline for many developing countries including China, India and the Like Minded Developing Countries group (LMDC)[m] since it would change the conditions of the article 4 of the Convention and the role of the Annex II Parties.

Examining the results of the negotiations at the financial field, the article 9 of the PA encompassed a shall provision in terms of the developed countries providing financial resources to assist developing countries with respect to both mitigation and adaptation in continuation of their existing obligations under the Convention (article 9.1). While, at the same time, it opened the door for other Parties to provide or continue to do it avoiding any new language since it is on a voluntary basis (article 9.2). It also recognised the significant role of the public sources of funding, but it also acknowledged that mobilisation from different sources is important, and that was the EU position in Paris as well as in previous meetings.

One of the most important conquers of the developing countries in the financial chapter, after losing the new financial goal, was the addition of articles 9.5 and 9.7 in terms of transparency and predictability of financial flows. The biggest opposition to these provisions did not come from the EU but from the Umbrella Group.

[l]As it was informed by European leader negotiators to the author during the different tracks of negotiations.

[m]More details about this group can be found in Bueno, 2018.

Other important components of the European position in Paris were the inclusion of an expert and non-political-based body to facilitate the implementation focused on mitigation commitments. In addition, the inclusion of carbon pricing and specific sectors in terms accountability such as LULUCF, aviation and maritime emissions. The mechanism to facilitate implementation and promote compliance was established under article 15 of the PA, consistent of an expert-based committee. Nonetheless, no particular sector was mentioned in the PA since the inclusion of different sectors represented a redline for many countries. In this regard, it is important to remember that the PA is a condensation of interests of maximum and minimum depending on the weight of the actors, and part of the diplomatic art developed by the French consisted in the clear identification of the redlines of all the groups.

The legal form was something critical from the very beginning, in particular, in terms of the real possibilities for the instrument to see the light in consideration of the 1997 Bird-Hagel Senate resolution that affirmed the United States would not ratify a climate agreement that threatens the American economy nor include emission reduction targets for major developing countries. During 2015, all the actors understood that the only way to avoid the US Senate was to forget the European idea of a protocol and to have an agreement that it was finally the annex of the Decision 1/CP.21.

The PA is an international agreement concluded between Parties in written form and governed by international law according to the Vienna Convention on the Law of Treaties, but it was adopted as an executive agreement with the only consent of President. The main misunderstanding in the legal nature of the agreement between the United States and the EU was the legally binding nature or not of commitments under the agreement. For the EU, as a consequence of its third-way vision, the mitigation commitments included in NDC shall be legally binding. For the United States, and according to its own bottom-up vision, the contributions do not need to be binding but a political aim of the countries.[n] This last interpretation was

[n]The legal advice of Daniel Bodansky on the legal nature of the agreement that was the position of the United States at a final stage of Paris is available in http://opiniojuris.org/2015/12/02/the-legal-character-of-the-paris-agreement-a-primer/, accessed in 14 January 2019.

even more evident with the withdrawal announced by President Trump in 2017. However, the measure will be effective in 2020.

Therefore, analyzing this critical moment in the climate change negotiations at the UNFCCC, we consider the European leadership was reaffirmed. In line with the vision, and this is probably the most critical point along with the distribution of efforts, even though negotiations' language seems to be clear in terms of the bottom-up approach of the PA based on the Chinese and the US interpretation, there are still some analysis pointing out its hybrid nature (Dimitrov, 2016; Costa, 2016; Parker et al., 2017) based on the transparency frameworks as well as the cycles. Moreover, it is still the official position and interpretation of the EU.

The internal coherence has been always the most important strength of the European climate policy, but it should be recognised that European NDC could have been more ambitious in light of the climate science and the temperature goal.[o] This feature is also applicable to the means of implementation provided to developing countries. If transparency is so critical for the European position, transparency of the support provided should also be developed accordingly. The different methodologies to define and classify climate finance inside the EU led to unfortunate scenarios with the OECD reports in relation to the 100-billion-a-year goal committed by developed countries in Cancun.[p]

According to the examination made of the main positions of the EU and the results obtained in the PA, it is possible to affirm that a large part of the positions were preserved and no redlines were crossed. Finally, the EU could continue its more pragmatic and flexible approach, in particular with developing countries and was patient with the Chinese–American alliance even when it did not respond to some of its critical interests. Parker, Karlsson and Hjerpe (2017) affirmed that Paris was a success for EU diplomacy. This can be explained by its instrumental leadership that included a continuation of its strategy of involving more with developing countries through

[o]For additional information on the fair share and ambition of EU NDC, visit Climate Action Tracker on https://climateactiontracker.org/countries/eu/

[p]Last OECD Report on climate finance is available at http://www.oecd.org/environment/cc/Climate-finance-from-developed-to-developing-countries-Public-flows-in-2013-17.pdf

informal coalitions that were inclined to avoiding the firewall, in this case, the High Ambition Coalition.[q]

Ultimately, according to the European decision-makers, the PA is not the solution for all our problems and climate change in general but a way to catalyse action at different levels and UNFCCC should continue to be a forum that regularly brings together evolving climate science with political leadership at the highest possible level (Lefevere et al., 2015).

The PA is not perfect; it was the achievable result with the circumstances available. It is weak in terms of adaptation, loss and damage, technology and emission reductions (Dimitrov, 2016) among many other issues. But nobody can deny it was a political success and that included the European diplomacy in a very special role.

2.3 The First Period of Implementation of the PA 2016–2018

After the adoption of the PA, an official signing ceremony was held in the UN Headquarters in New York on 22 April 2016. Considering the political context in the United States and the Presidential campaign, the UN Secretary General, Ban Ki-moon, along with many leaders, and in particular, Obama, as well as the Chinese and the EU, promoted an accelerated entry into force of the PA. The goal was to achieve that the Agreement could enter into force before the COP22 in Marrakesh and the Presidential elections in the United States, compelling the new administration to remain in the Agreement until 2020 according to the withdrawal clauses. The strategy was effective since the PA entered into force on 4 November 2016 and the Trump administration announced the withdrawal on June 2017. Until today, the PA has 195 signatories and 184 parties and the number continues growing.[r]

The Paris Outcome included Decision 1/CP.21 and the PA as an annex to the decision. This decision is very important to understand the first period

[q]The High Ambition Coalition was originally composed in by Angola, Marshall Islands, Germany, Grenada, Peru, Santa Lucia, UK, Gambia, Colombia, Chili, Mexico and Switzerland. During 2015 and after Paris, more than 80 Countries joined.

[r]Official information on the PA ratification status is available in the UN Treaty Collection web page at https://treaties.un.org/Pages/ViewDetails.aspx?src=TREATY&mtdsg_no=XXVII-7-d&chapter=27&clang=_en

of implementation that was originally planned after 2020 and for all the reasons aforementioned, it took place from 2016 to 2018 being conducted by the APA and the subsidiary bodies of the Convention, the Subsidiary Body for Implementation (SBI) and the Subsidiary Body for Scientific and Technological Advice (SBSTA).

The struggle started since the very beginning with the negotiation of the APA agenda at the subsidiary bodies meeting celebrated in Bonn in May 2016. It took more than half of a meeting the negotiation for the adoption of the APA agenda and its final format excluded several elements of the PA. Given that the excluded elements concerned the needs and interests of the developing countries, such as means of implementation and loss and damage, among others, developing countries agreed with developed countries that the agenda item 8 had a more open nature through the language "further matters." In any case, what elements to include in this point was a new battle for each APA meeting.

For the EU delegation, the most important issues to include in the APA agenda were mitigation, transparency, compliance and the(GST). All of them were incorporated with the divergences on the nature of agenda item 3: "further guidance in relation to the mitigation section of decision 1/CP.21." Summarising, the different interpretations in the chapeau referred to the nature of article 4 of the PA that, in the understanding of the Chinese and the LMDC group, is not only about mitigation since headings of the Agreement were removed, differently to Decision 1/CP.21 where headings remained. So that, the objective of the negotiations was to find a language that was respectful with this two interpretation of article 4, but at the same time served its purpose and the purpose was to develop further guidance for the mitigation component of the NDC.

After the adoption of the agenda, the key task was the development of additional aspects for the submission of the mitigation component of the NDC while respecting the nationally determined nature. In that sense, the debate on the top-down and the bottom-up visions continued to be at the center of the dispute as more clarity was required to aggregate efforts more effectively.[s] But at the same time, what gases to include,

[s]As it resulted of the synthesis reports of the aggregated effect of INDC prepared by UNFCCC Secretariat in 2015 and 2016 (UNFCCC, 2015; 2016).

target type, target year, reference point(s), timeframes, scope and coverage, how to show that the contribution is fair and ambitious and various other aspects, governed the debate before and after Paris to Katowice.

In terms of the other agenda items, the EU was flexible with the inclusion of adaptation in the agenda item 4 referred: further guidance of the adaptation communication. Anyway, it is clear that the first victories in terms of the interests of the EU came with the adoption of the agenda where the position was well-preserved.

One more time after Morocco (COP22) and Fiji (COP23) Presidencies, the EU hosted an international conference where it was supposed to achieve a new milestone in the UNFCCC negotiations, that it was to complete the Paris Work Programme. The Polish hosted its third COP in the city of Katowice with a carbon history.

Based on the official documents, the EU position and priority to Katowice was grounded on the urgent need to adopt a strong Paris Work Programme with clear provisions, minimally in transparency, finance, mitigation and adaptation. The outcome had to preserve the spirit of the PA, be applicable to all Parties, take into account different national circumstances and reflect the highest possible ambition over time (European Commission, 2018).

The outcome in Katowice included all the agenda items under the APA and the subsidiary bodies, but the only topic that was postponed is a decision on article 6 including markets and non-markets mechanisms. This topic was critical for the EU, as well as many other actors, including Switzerland. However, the level of maturity of the proposals was not enough to achieve, at least, one acceptable basis for all actors as a starting point. And even though this topic delayed the closing of the COP one more day, the EU knew a long time ago that a result in Poland would not be achieved in this respect and that given the risks of losing all the package, the pragmatic position was the best solution, and it was.

Giving more details to the positions and the results achieved, in transparency—a big priority of the European team—the modalities, procedures and guidelines (MPG) of the enhanced transparency framework for action and support were finally adopted based on the biennial transparency reports (BTR), the technical experts review (TER) and the facilitative multilateral consideration of progress (FMCP) and using as starting point the "should" and "shall" provisions of the PA.[t]

The purpose of the enhanced transparency framework was to provide a clear understanding of climate change action based on mitigation, adaptation, finance, technology and capacity-building activities. The guiding principles of the MPG included the special circumstances of LDC and SIDS, the improvement over time, the flexibility to developing countries that need in light of their capacities, avoid duplication of work, double counting and burden, ensure frequency of reporting, promote accuracy, completeness, consistency and comparability and ensure environmental integrity. So far, all ideas were acceptable for the EU position even if self-determination of the flexibilities was a struggle of three years successfully solved in favor of the developing countries.

It is also worth mentioning that the MPG include a loss and damage section, which was acceptable for the EU. However, it was strongly rejected by the Umbrella Group. One of the most controversial sections was the transparency of support provided, where the result was accurate in terms of maintaining the balance of the package. It also included provisions for the support received by developed countries and even though it was a should provision and will require a lot of practice in developing countries, it is important for the cycle of transparency and the accountability of the resources that developed countries will communicate as provided. Not only this, but it also included clauses on the support needed by developing countries that was something very resisted by developed countries.

If the objective was to provide incentives to build domestic institutions, data collection and tracking systems as announced by the EU, the MPG were according to its expectations.

In terms of climate finance, the EU knew that this was the angular piece to unlock all the package from the developing countries perspective and

---

[t]This means that the submission of national inventories as well as information to track progress in relation to NDC under article 4 are "shall" provisions for all Parties while adaptation is a should provision. For that reason, the BTR shall include the first two components with the special circumstances of LDC and SIDS and the self-determined flexibilities of all developing countries. However, the BTR may include adaptation if the Party decide to do so and there are possibilities to merge documents with the further guidance of the adaptation communication. In addition, developed countries shall provide information in their BTR in relation to the provision of means of implementation to developing countries (finance, technology and capacity building). The TER only refers to "shall" provisions.

considering the backtracking of the United States. In consideration of that, the EU supported the Adaptation Fund (AF) serving the PA after many years of strong discussions about the changes that the donors demanded, starting with the Board Composition. However, the final decision on the AF was lighter than thought and nearer to the G77 and China demands in Paris and Marrakesh for a procedural decision.

The EU also supported the dialogue on the long-term climate finance, but all the developed countries jointly blocked any possibility to achieve a new financial goal after 2020. The decision on this matter was postponed after 2020.

Additionally, and with big opposition of developed countries, but in particular of the Umbrella Group, the modalities according to the article 9.5 of the PA came out.[u]

The official position of the EU specified that the EU and its Member States were the biggest donors of climate finance to developing countries and that they were providing their fair share of the overall goal of 100 billion per year to 2020. This is part of its leadership feature concentrated on internal coherence. A similar idea applied to the adaptation agenda items, where the EU tried to show, not only their support to the international dimension of adaptation based on cooperation, lessons learned and sharing experiences, but also, its own work at home. This includes the EU Adaptation Strategy in 2013 and the last report of the Commission evaluating the strategy, as well as the increasing number of national, regional and local adaptation strategies in the Member States (European Commission, 2018).

The adaptation agenda is crosscutting in terms of the different sections of the PA and the results coming from COP24. Nevertheless, two of the most important outcomes were the further guidance for the adaptation communication and the implementation of paragraphs 41, 42 and 45 of the Decision 1/CP.21 that had the recommendations of the Adaptation

[u]The article 9.5 of the PA established the compromise of the developed countries (shall provision) to biennially communicate indicative quantitative and qualitative information on public finance to assist developing countries for both mitigation and adaptation actions, as well as resources coming of private sources and other sources.

Committee, the Expert Group of the LDCs and the Standing Committee on Finance.[V]

The further guidance in relation to the adaptation communication was a short document very specific and flexible for everyone. It determined that the adaptation communication purpose was to increase the visibility of adaptation, strengthen adaptation action and support for developing countries, enhance learning and understanding of adaptation needs and actions and provide input to the GST.

It maintained the flexibility under article 7.11 in terms of the document that can officiate as the context for the submission of the adaptation communication and develops a list of elements with some distinctions according to the G77 and China proposals. It included some weak clauses in terms of support and determined that the Adaptation Committee would develop supplementary guidance with the engagement of the IPCC Working Group II by 2022.

The EU was more flexible with the negotiation of the further guidance in relation to the adaptation communication than with the paragraphs 41, 42 and 45. Probably, as a result of the risks in the second case, including: the mobilisation of support and other methodological issues, the relation with the GST and the recognition of efforts. The EU did not have a leading role in adaptation negotiations what facilitated that the final outcomes were closer to the G77 and China positions. Anyway, it should be recognised that the EU did not block adaptation negotiations but tried to have a facilitative role.

The mitigation outcomes, in particular, the further guidance to the mitigation section of decision 1/CP.21 on features, information to facilitate the clarity, transparency and understanding and accounting of NDC were possible even when it was one of the most controversial. The body of the CMA decision was full of recalls with not many decisions. In terms of features, the proposal of Brazil that then turned to Argentina, Brazil and Uruguay

[V]The mandates under paragraphs 41, 42 and 45 of Decision 1/CP.21 included ways to enhance the coherence of the work of adaptation-related institutional arrangements; modalities for recognizing the adaptation efforts of developing country parties; methodologies for assessing adaptation needs with a view to assisting developing countries without placing undue burden on them; methodologies for taking the necessary steps to facilitate the mobilisation of support for adaptation in developing countries and methodologies for reviewing the adequacy and effectiveness of adaptation and support.

(ABU) proposal was accepted. It affirmed that Article 4 of the PA already included the features of the mitigation component of the NDC. The most substantive results were in information and accounting under annexes I and II. For information, it included quantifiable information on the reference point; time frames and/or periods for implementation; scope and coverage; planning processes; assumptions and methodological approaches; how the Party considers that its NDC is fair and ambitious in the light of its national circumstances and how it contributes towards achieving the objective of the Convention.

In relation to accounting, it comprised common metrics and methodologies of the IPCC, seeking for methodological consistency between communication and implementation of NDC; the inclusion of all categories of anthropogenic emissions or removals and the need to provide an explanation if a category is excluded.

The information would be applicable to the second and subsequent NDC, but it encouraged Parties to apply it for the first contribution by providing a second due date that would be 2020 and that was very critical for the EU. The EU had also proposed more details in terms of the information, accounting and features, but the result could be considered as balanced bearing in mind the different starting points and the differentiated responsibilities and capacities.

Another issue that can be mentioned is that the CMA decided to account the NDC in the BTRunder the transparency framework including a structured summary. This summary was pursued by the EU position.

Using the strategy of "leadership by example" (Parker and Karlsson, 2017) and seeking of the traditional internal coherence of that leadership, the European Commission presented a vision based on achieving climate neutrality by 2050. Moreover, the European Commission (2018) before COP24 announced that the EU was well on course to exceed its 2020 target and working actively for the NDC commitment. The new actions encompassed the establishment of an economy-wide framework of legislation based on the modernisation of the EU ETS; new energy efficiency and renewable energy targets; new 2030 targets for all Member States to reduce emissions in non-ETS sectors including transport, buildings, agriculture and waste.

Decisions have been also taken in the GST, the Committee to facilitate implementation and promote compliance, public registries, response measures and common timeframes, among many others.[w]

Considering the process and the outcomes in light of the EU positions and the features of its leadership in climate change negotiations as proposed, we interpret flexibility and the adjustment to changing circumstances was the most important characteristic of the Polish Presidency during this COP as well as of the EU during this period. Throughout 2018, it often seemed that a result in Katowice would be impossible for many reasons and especially the stagnation in terms of climate finance, which also helped to paralyse the already difficult mitigation and transparency talks. The slogan of no more obligations and burdens for developing countries without means of implementation was evident in this process.

It was already clear before that the vision was not shared and only the EU, with some allies, interpreted the regime in a hybrid matrix. A distribution of mitigation efforts was impossible after Paris where the Copenhagen designed prevailed. Nevertheless, an enhanced transparency framework was achievable with the flexibility of developed countries in some very strategic points: transparency of support including article 9.5[x]; self-determined flexibility; the methodological and technical strengthening of the mitigation component of the NDC without strangulating its nationally determined nature and some political signals in terms of climate finance.

Given that, the EU continued to be an important actor in climate change negotiation during the first period of implementation of the PA, but many concessions were necessary, not only in favor of China and the United States but also for other groups of developing countries such as the African Group.

Conclusion

This contribution aimed to analyse the changes experienced in EU leadership in three moments from 2009 to 2018. It started with the Copenhagen failure where the EU seemed to lose its leadership qualities and ended with

[w] COP24 outcomes are available in https://unfccc.int/katowice

[x] Without a result on the implementation of the article 9.5 of the PA, the African Group of Negotiations would not have accepted the final outcome of COP24.

2018 as the closure of the first period of implementation of the PA during COP24.

Considering the vast trajectory of leadership literacy applied to this topic, the contribution underscored four features of EU leadership at the international climate change arena including its vision; the internal coherence; the degree of achievement of its objectives and proposals and the flexibility to adapt to a changing context.

The first period analysed that covered Copenhagen to Cancun Conferences could be characterised as a journey starting with a lost but culminating with a recovery of leadership based on the delicate balance of recognizing the mistakes and the lack of flexibility in the position and the strategy during COP15. In addition, the EU understood that leading by example and internal coherence feature could be an asset or a burden depending on how to use it as a negotiation bargaining. This knowledge was applicable to the three moments, including a more cooperative approach with the developing countries and its different groups and constituencies.

To achieve the PA was necessary to assume a pragmatic approach to the vision by leaving the door open for different interpretations. The EU maintained that the Paris architecture was hybrid, whereas the United States and the Chinese along with many others were strong in the bottom-up approach perspective anchored in the nationally determined nature of the NDC as the vehicle for universal climate action in the PA.

After adopting a pragmatic and more realistic approach, as well as, a more cooperative one, the degree of achievements improved. Considering the negotiations through a prism of zero-sum could not and cannot be beneficial for the interests of the EU. Letting go some battles can help to obtain structural gains much more beneficial in terms of its international leadership, but also of the distribution of internal efforts that have managed to organise even with the changes of international conditions.

The learning by doing process will continue and the challenges to the leadership are a lot and will be more in a changing environment with the US withdrawal and its serious implications in terms of mitigation and finance.

A new phase will start in 2019 with no political bodies to guide decisions in the UNFCCC outside the subsidiary bodies. After many years of many ad hoc working groups, at least for now, the Convention will open a process of implementation based on results and those results should be those of the

Parties. For that reason, the technical inputs of the agency will be even more important for facilitating the work, especially of the developing countries.

References

Afionis, S. (2011) The European union as a negotiator in the international climate change regime. *International Environmental Agreements: Politics, Law and Economics*, 11(4), 341–360.

Andresen, S. and Agrawala, S. (2002) Leaders, pushers and laggards in the making of the climate change. *Global Environmental Change*, 12, 41–51.

Bäckstrand, K. and Elgström, O. (2013) The EU's role in climate change negotiations: From leader to 'leadiator'. *Journal of European Public Policy*, 20(10), 1369–1386.

Bodansky, D. (2010) The Copenhagen climate change conference: A postmortem. *American Journal of International Law*, 104(2), 230–240.

Bodansky, D. (2011) A tale of two architectures: The once and future UN climate change regime. *Arizona State Law Journal*, 43, 697-713.

Bueno, M.P. (2014) La construcción de políticas climáticas europeas y su Internacionalización: Desafíos pasados, actuales y futuros hacia 2020. *Cuadernos Europeos de Deusto*, 51, 123–153.

Bueno, M.P. (2017) El Acuerdo de París: ¿una nueva idea sobre la arquitectura climática internacional? *Revista Relaciones Internacionales*, 33, 75–95.

Bueno, M.P. (2018) Cambio, identidades e intereses: Argentina en las negociaciones multilaterales de cambio climático 2015–2017. *Colombia Internacional*, 96, 115–145.

Compston, H. and Bailey, I. (eds.) (2008) *Turning Down the Heat. The Politics of Climate Policy in Affluent Democracies*. Palgrave Macmillan: Basingstoke.

Costa, O. (2016) EU policy responses to a shifting multilateral system, in Barbé, E.,Costa, O. and Kissack, R. (eds.), *Chapter 6 Beijing After Kyoto? The EU and the New Climate in Climate Negotiations*. Palgrave Macmillan: London, pp. 115–133.

Curtin, J. (2010) *The Copenhagen Conference: How Should the EU Respond?* Institute of International and European Affairs, Dublin.

Damro, C. and MacKenzie, D. (2008) Turning down the heat. The politics of climate policy in affluent democracies, in Compston, H. and Bailey, I. (eds.), *Chapter 4, The European Union and the Politics of Multi-Level Climate Governance*. Palgrave Macmillan: Basingstoke, pp. 65–84.

Delbeke, J. and Vis, P. (eds.) (2015) *EU Climate Policy Explained*. Routledge: New York.

Dimitrov, R.S. (2016) The Paris agreement on climate change: Behind closed doors. *Global Environmental Politics*, 16(3), 1–11.

European Commission (2009) *IP/09/1867*, 2 December 2009, available at: http://europa.eu/rapid/pressReleasesAction.do?reference=IP/09/1867

European Commission (2011) *Memo/11/825*, 24 November 2011, available at: http://europa.eu/rapid/press-release_MEMO-11-825_en.htm

European Commission (2018) *Memo/12/6592*, 3 December 2018, available at: http://europa.eu/rapid/press-release_MEMO-18-6592_en.htm

European Council (1990) *Dublin Summit 1990*, 25–26 June 1990, available at: http://www.europarl.europa.eu/summits/dublin/default_en.htm

European Council (2015a) *Council Conclusions on Climate Finance. Press Release 797/15, Press Release 797/15*, 11 November 2015, available at: https://www.consilium.europa.eu/en/press/press-releases/2015/11/10/conclusions-climate-finance/pdf

European Council (2015b) *EU Position for the UN Climate Change Conference in Paris: Council Conclusions. Press Release 657/15*, 18 September 2015, available at: https://www.consilium.europa.eu/en/press/press-releases/2015/09/18/conclusions-un-climate-change-conference-paris-2015/pdf

Groen, L., Niemann, A. and Oberthür, S. (2012) The EU as a global leader? The Copenhagen and Cancun UN climate change negotiations. *Journal of Contemporary European Research*, 8(2), 173–191.

Groenleer, M.L.P. and Van Schaik, L.G. (2007) United we stand? The European union's international actorness in the cases of the international criminal court and the Kyoto protocol. *Journal of Common Market Studies*, 45(5), 969–998.

Gupta, J. and Grubb, M. (eds.) (2000) *Climate Change and European Leadership. A Sustainable role for Europe?, Springer Nature: Switzerland.*

Gupta, J. and Ringius, L. (2001) The EU's climate leadership: Reconciling ambition and reality. *International Environmental Agreements: Politics, Law and Economics*, 1(2), 281–299.

IPCC (2007) Climate change 2007: Synthesis Report. Contribution of Working Groups I, II and III to the Fourth Assessment Report of the Intergovernmental Panel on Climate Change (IPCC Geneva, Switzerland).

Lefevere, J., Runge-Metzger, A. and Werksman, J. (2015) EU climate policy explained, in Delbeke, J. and Vis, P. (eds.), *Chapter 5 The EU and International Climate Change Policy, en.* Routledge: New York, pp. 94–108.

Oberthür, S. (2009) The European union and global governance, in Tellò, M. (ed.), *Chapter 9 The Role of the EU in Global Environmental and Climate Governance.* Routledge: London, pp. 192–209.

Oberthür, S. (2011a) The European union's performance in the international climate change regime. *Journal of European Integration*, 33(6), 667–682.

Oberthür, S. (2011b) Global climate governance after Cancun: Options for EU leadership. *The International Spectator*, 46(1), 5–13.

Parker, C.F. and Karlsson, C. (2017) The European union as a global climate leader: Confronting aspiration with evidence. *International Environmental Agreements: Politics, Law and Economics*, 17(4), 445–461.

Parker, C.F., Karlsson, C. and Hjerpe, M. (2017) Assessing the European union's global climate change leadership: From Copenhagen to the Paris agreement. *Journal of European Integration*, 39(2), 239–252.

Parker, C., Karlsson, C., Hjerpe, M. and Linnér, B.O. (2012) Fragmented climate change leadership. Making sense of the ambiguous COP-15 outcome. *Environmental Politics*, 21(2), 268–286.

Schreurs, M. and Tiberghien, Y. (2007) Multi-level reinforcement: Explaining European union leadership in climate change mitigation. *Global Environmental Politics*, 7(4), 19–46.

UNFCCC (2015) Synthesis report on the aggregate effect of the intended nationally determined contributions. United Nations Framework Convention on Climate Change, available at: https://unfccc.int/resource/docs/2015/cop21/eng/07.pdf

UNFCCC (2016) Aggregate effect of the intended nationally determined contributions: an update, United Nations Framework Convention on Climate Change, available at: https://unfccc.int/resource/docs/2016/cop22/eng/02.pdf

Van Schaik, L.G. and Schunz, S. (2012) Explaining EU activism and impact in global climate politics: Is the union a norm- or interest-driven actor? *Journal of Common Market Studies*, 50(1), 169–186.

Verolme, H. (2012) European climate leadership Durban and beyond, Discussion Paper, (Heinrich Böll Stiftung Brussels).

Vogler, J. (1999) The European union as an actor in international environmental politics. *Environmental Politics*, 8(3), 24–48.

Vogler, J. and Bretherton, C. (2006) The European union as a protagonist to the united states on climate change. *International Studies Perspectives*, 7, 1–22.

Wu, F. (2012) Sino-Indian climate cooperation: Implications for the international climate change regime. *Journal of Contemporary China*, 21(77), 827–843.

Wurzel, R. and Connelly, J. (eds.) (2010) *The European Union as a Leader in International Climate Change Politics*. Routledge: New York.

Yamin, F. (2000) Climate change and European leadership: A sustainable role for Europe?, in Gupta, J. and Grubb, M. (eds.), *Chapter 4 The Role of EU in Climate Change Negotiations. Springer Nature: Switzerland*, pp. 47–66.

Zito, A.R. (2005) The European union as an environmental leader in a global environment. *Globalisations*, 2(3), 363–375.

CHAPTER 3

Evolution and Overview of the Current European Legal Commitment Heading for a Unique Model of Sustainable Development

María Dolores Sánchez Galera

The evolution of EU environmental law and governance has taken the lead towards the emergence of a unique EU sustainability model with its imperfections and challenges, but with greedy ambitions within the global scenario.

Introduction

This contribution aims to approach the most prominent aspects of the EU unique model of sustainable development in which climate change policy, strong environmental protection and environmental procedural rights, and sustainable economic growth embracing important social policy and welfare aspects are at the forefront of every other sectorial policy. The sustainability paradigm on which the current EU policy horizon relies on has been nurtured on two aspects: (1) the evolution of a strong integrated environmental protection policy at EU level and (2) the global governance context and interaction among many legal orders that has determined today's constitutional existence of sustainability in the EU. Theoretically, the different aspects of this so-called Anthropocene era are at stake in social science discourses in order to implement laws and public policies capable of providing the right protection and implementation mechanisms for sustainable development in today's societies.[a]

Thus, the present work is structured along two main parts. The first part focuses on the content of two major new paradigms in legal sciences: global governance and sustainability. We focus on an analysis of the normative framework of the concept of "sustainability" and "sustainable development" as well as the consequences of multilevel governance at the international level. Attention will be drawn to global environmental law within the wider context of the emergence of the so-called "global law." The author believes that the different layers of governance at the global level determine the grounds to understand the evolutions of environmental protection beyond the state boundaries and suggest the roots of the "normative" framework of sustainable development at regional, national and local levels today. The second part of the contribution will present the constitutional basis and political grounds to interpret the legal settings of "sustainability" in the European context and its major challenges.

3.1 A Brief History of Sustainability: A Conceptual Approach to "Sustainability" and "Sustainable Development"

Today those environmental threats are arising from several fronts, the paradigm of "sustainable development" is enshrined by a growing number of international treaties, constitutions, and agreements. "Sustainable development" could be said as the new paradigm of our era; potentially it could embrace indistinctively every field of action. As (Sachs, 2015) puts, it is a "way of understanding the world and a method for solving global problems." Still, the idea of sustainability is as old as human existence (Caradonna, 2014). In fact, before engaging further with the origins of the sustainability concept, it is important to devote some introductory comments on the

[a]The contemporary debate about nature and the human influence in relation to nature has been marked by the development of a geological hypothesis that has rapidly gained steam outside the natural sciences realm, namely, that of the Anthropocene. According to the "Anthropocene" the Earth's physical change (i.e., biodiversity, climate, and so on) is mainly driven by human activity. We are, of course, the main drivers of our global-scale environmental disorder. For further analyses on this topic, see Arias-Maldonado, M. (2015) *Environment and Society: Socionatural relations in the Anthropocene*, Springer.

differentiation between "sustainability" and "sustainable development." Both terms are today used interchangeably in general by the legal scholarship, although they have distinct meanings.

"Sustainability" precedes sustainable development; it pre-dates the late-20th-century concept of "sustainable development" (Bosselmann, 2010). "Sustainability" focuses on the "capacity for humans to live within environmental constraints" (Robinson, 2004) and the heart of this concept is the respect for ecological limits and ecological integrity. It is important to remark that sustainability pre-dates the normative character of "Sustainable development." Its existence belongs to our era. Regarding the historical evolution of the term "sustainability," the term emerged first during the age of enlightenment. The backdrop against which the term sustainability was coined was again an ecological crisis consisting of deforestation caused by rapidly increased economic demands.[b] Knowingly, the concept was given a name in the early 18th century by a Saxon bureaucrat Hans Carl von Carlowitz working in the mining sector who coined the term *Nachhalligkeit* to describe the practice of harvesting timber continuously from the same forest.[c] But, indeed, sustained yield forestry took shape at this time not only in Western Europe but also in Japan, other parts of Asia and on Colonial islands in both the West and East Indies (Caradonna, 2014). Thus, it is among 19th-century theorists of forest management that "sustaining" life as a whole to provide for human life was a commonly shared view. This view was confined to scholars and forest academies in Germany, but forest academies elsewhere followed the same ecological context. So, during the 19th century, the concept emerged as the central term within forest science growing broadly to include the entire spectrum of the ecosystem "forest" comprising locations, soil fertility, organism diversity, wildlife habitat, water reservoir, protection against erosion, "lung" function, and

[b] As Bosselmann explains quoting the Works of German and British naturalists, "[b]y 1650 widespread shortages of wood began to cripple the economies in European countries. At the same time, the new discipline of forest science and management emerged. Its focus was on studying the conditions for sustained forestry and sustainable yield" (Bosselman, 2008); with further reference to the works of Radkau, J, *Natur und Macht. Eine Weltgeschichte der Umwelt* (Evelyn, 1664).

[c] Carlowitz learned about the dependence of mining on its natural resource base and a year before his death he published his work summarising these professional and lifetime experiences. See further Carlowitz (1713).

recreational space.[d] These were the first traces of a holistic approach of sustainability having at its core a strong "ecological content" and giving the right guiding tools to the forestry managers of that time.

"Sustainability" goes beyond the main difficulties faced to date in environmental protection. Its existence focuses on action as opposed to the defensive characteristic that environmental law and environmental measures bear. It is itself rooted in the concrete ecological boundaries of sustainable development. "Sustainable development" instead has achieved a normative stance on the world, putting forward an intergenerational concept of development. The parameters of "sustainable development" launched by the international community as a whole currently fulfil a chief ecological content that encompasses the right of future generations and social justice as having a specific and prominent role for policy guidance in every action, including justice, equity, education, poverty eradication, culture, urban development and safety among other global values.

3.2 Sustainable Development Today as a Multidimensional Concept

"Sustainable development" today is one of the basic political objectives of the UN, together with peace, international security, and human right protection. This global policy reached its glorious moment on September 25, 2015, with the document "Transforming our world: Agenda 2030 for Sustainable Development." The countries of the world adopted a set of goals (17 development goals) to end poverty, protect the planet and ensure prosperity for all humanity as part of a new sustainable development agenda. A key element of these 17 goals (sustainable development goals (SDGs)) with targets to be achieved over the next 15 years is that everyone needs to contribute: governments, private sector, civil society and individuals.[e] These 17 objectives take inspiration from the Millennium development goals, but their most remarkable difference is that they were formulated by

---

[d]Other forest academies that followed the same German view of sustainability were Austria-Hungary, Switzerland, France, Russia, Scandinavia, the United Kingdom with its colonies and the United States of America. Confr. Bosselman (2008) referring to the works of Ulrich Grober, in his analysis of the terminological history of sustainability (Grober, 2007).

[e]Available at: http://www.un.org/sustainabledevelopment/ (last time visited on 12 April 2019).

developed countries. Against this background, SDGs are universal despite their differing applications.

The present evolution takes us to the (still) unresolved (but also quite unimportant regarding its current practical and global policy dimension) question of the "multidimensionality" of the notion of "sustainable development." In fact, during its recent progression, we understand that it can be to a political objective, a legal concept, a human right, a methodological framework for creating and applying of public policies and international norms or even a hermeneutical instrument for legal interpretation. Rodrigo points even to its social dimension as an open-wide prerequisite of development accompanied by dimensions of peace, security, and culture (Rodrigo, 2015).

Even for those who embrace sustainable development in its most accepted three-partite model including environmental, social, and economic components, it is impossible to ignore that "sustainable development encompasses the development and application of the concept of sustainability in a broad range of domains: urbanism, agriculture and ecological design, forestry, fisheries, economics, trade, population, housing and architecture, transportation, business, education, social justice and so on" (Caradonna, 2014). Therefore, the components are not just three. We should be aware that the more recent intervention and intertwined connections of pure science in other scientific domains have revisited the notion of sustainability and have helped climate change policy to emerge in the global political debate at all levels. Furthermore, effective governance is a key component that renders sustainable development the guidance of future global governance. Good governance is evidently an important component of sustainability. It includes the institutional parties, private actors, stakeholders, and associations or political components at every level of governance. It was strongly proclaimed by the already mentioned Delhi Declaration (2002) by ILA, and it has subsequently been proclaimed by the United Nations (2014). This multidimensional nature of the concept has helped national and regional governments to adopt different policies to strong multifaceted governance. "Good governance" as a political objective has been enshrined in the Agenda 2030 for the Sustainable Development of the UN as well as at the regional level in the European Union (EU). It does not refer to governments, and it relates to global private actors, stakeholders, and multinationals acting in the global scenario and it affects our living world. The current challenge is the fragmented existence of good governance in a world aiming at not giving up sovereign control of its natural resources regardless of management types, with major social actors of the business world running after their own share. We are running the risk of getting into what it has been called "green developmentalism" (McAfee, 1999) that strives for market-oriented solutions and technological development that leaves untouched the current models of production and governance in our current global legal order.

Here, it becomes important to ascertain the legal dimension of the concept to identify the interpretative functions of sustainable development that could fit within traditional categories of normativity in the international arena (Barral, 2012). In fact, according to Bosselmann, sustainability as a norm "can be formulated as the obligation to promote long-term economic prosperity and social justice within the limits of ecological sustainability" (Bosselmann, 2008); therefore, according to this, sustainability would slip into the lands of legal principles getting normative quality even though it has not yet been recognised as such by international law (Lowe, 1999). This line of reasoning is also followed by more recent scholarship, considering that the normative value of sustainable development could have a myriad of expressions. Therefore, sharing the aforementioned view on the concept of sustainable development, we agree with the most authoritarian scholarship that "the concept of sustainable development is located somewhere in the spectrum of (non-legal) policies and (legal) norms" (Lowe, 1999). We can perhaps exclude two extremes. On the one hand, the concept is more than a mere political ideal as it has been referred to repeatedly in hard law (i.e. Climate Change Convention 1992, Article 3; Biodiversity Convention 1992, Articles 8, 10; Desertification Convention 1997, Articles, 4, 5; Antigua Convention 1992, Article 2)[f] and in soft law.

On the other hand, the concept is less than a rule as it lacks direct legal consequences. "The proximity to policies seems closer than to a rule, but how close? This depends on the degree of political and legal guidance that the concept can provide." (Bosselmann, 2008) This aspect is paradigmatic for International law, namely the case concerning the

[f]For further examples see, Cordonier Segger and Khalfan (2004).

Gabčikovo-Nagymaros Dam (Hungary/Slovakia)[g]; in this case, the International Court of Justice (ICJ) used 'sustainable development' as the argument to resolve the case, but with such "caution and delicate ambiguity in the phrasing of the passage" that there is no a well-established shift towards the belief that "sustainable development" could always become an element of hermeneutic interpretation for the court in similar conflicts, although later interpretation of the ICJ has followed that road.[h] The relevance of this judgment is that there is room to follow the significant flow of acceptance towards the establishment of sustainable development as a binding legal norm in the international community—more as a political objective than a legal principle (Lowe, 1999).

3.3 Some Notes on Global Governance and Sustainability: Global Environmental Governance

In the fast-changing global landscape, we are confronted with global-scale environmental disruption. Global governance that promotes sustainable development is one of the big challenges faced by political decision-makers worldwide and, of course, in Europe. In fact, the concept of "global environmental law" is increasingly used to challenge the inter-state paradigm of international environmental law, and it has determined a focus on issues of common interest to humanity as a whole (Morgera, 2012; Yang & Percival, 2009; Percival and Piermattei, 2014; Esty and Ivanova, 2004), such as sustainable development and intergenerational justice. This is where the complex global governance of our time comes about. Within this context "Sustainability" emerges as an inclusive, flexible, and evasive

[g]*Gabčikovo-Nagymaros Dam Case (1997) ICJ Reports 7, para. 140.* The case arose out of a 1977 Treaty in which Hungary and Czechoslovakia (as it then was) agreed on a joint project to build hydroelectric facilities and improve navigation and flood control on the Danube. In 1989, Hungary suspended, and later abandoned, work on the agreed project because, it said, of serious criticisms of the environmental impact of the project. The two states concluded in 1993 a special agreement for the reference of the dispute to the International Court. The important thing about this judgement is that Judge Weeramantry on his judgement distinguished his conception of sustainable development from that adopted by the majority in the Court: "...I consider it to be more than a mere concept, [it is]...." a principle with normative value. For a better understanding of the case, see further Lowe (1999).

[h]*Iron Rhine Railway Railway Arbitration* (2005) PCA, paras. 58–59; *Pulp Mills Case* (Provisional Measures) (Argentina v Uruguay) (2006) ICJ Reports, para. 80; *Pulp Mills Case* (Merits) (2010) ICJ Reports, para. 177.

concept, even though it is not clearly classified. Nevertheless, the concept is important enough to guide international-law making and the growing role of global institutions. European governance is following the trait of protecting the common interest of humanity through a complex normative and political package addressing energy, climate protection, and sustainable governance as a whole. Thus, the EU is going beyond the past and current achievements of the integration principle that played the most significant role in the evolution of EU environmental law and governance. It aims to integrate environmental policy in other sectoral policies to tackle intergenerational prevarication today as it has originally been conceived at the international level.

Global acceptance of the term following the different international conventions has welcomed what some authors have called a "Global Policy on Development" (p. 36). Evidence of such successful acceptance is not only represented by the most recently launched 17 SDG, but different conventions such as the United Nation Convention on Climate Change (UNCCC), Kyoto Protocol and the Convention on Biological Diversity are also good examples of this. These international conventions show that regardless of the challenges that the object of these cooperation agreements represents the international and global scenario, a better framework for action under the auspices of social justice and equity is already being developed. Such agreements provide an illustration of the importance of globally inclusive regulatory regimes not guided by (legally) recognised legal principles such as the "principle of sustainability." The other important evidence is brought by Agenda 21, COP 21, Benefit Sharing Mechanism regulated by the Biological Diversity Convention and public participation in regulatory processes. Furthermore, because there is the case law of the ICJ or the Organ of Solution of Differences of the WTO, institutions have invoked it for the interpretation of international norms (Rodrigo, 2015). UN has now placed the SDG among the sacred traditional UN objectives of Peace and Security.

As a result, more than ever, it is necessary to face the existence of a global order, its evolving nature, and the interaction between many legal orders. Additionally, more implementation instruments and further accountability for effective functioning are needed. If we compare international environmental law with other international law areas (i.e., human rights law, international labour law, or international trade law), we can see that is clearly underdeveloped. It has no globally binding instrument that sets out the rights and duties of states with respect to the environment. Still, the absence of a foundational treaty to protect the global environment has not determined the failure of the integrative processes of ecological concerns and the arrival of an effective global environmental law scenario. Things are slowly changing, and in 2017, the Global Pact for the Environment was launched as an initiative to conclude a legally binding international instrument under the United Nation 72/277 entitled "Towards a Global Pact for the Environment." This initiative aims to provide an overarching framework to the international environmental law for further solidifying, consolidating, and advancing international environmental law by considering pressing environmental challenges. The initiative also seeks to improve the implementation of international environmental law as part of the spirit of compliance with the SDG as well as the globally agreed environmental goals and targets.

Despite the latest efforts to enforce a global pact for the environment, by contrast to the little material existence of a body of binding international environmental law, an emergent branch of lawyers is building on theoretical discourses about the existence of global constitutional and administrative law[i] for the importance of regulatory governance to implement mechanisms of accountability, participation, and responsiveness (Stewart, 2014). According to Morgera, "[T]he concept of global environmental law assist in understanding the 'functional' role of states and the 'functionalization of national sovereignty' arising from the evolution of international environmental law in the context of the plurality of legal orders. States exercise 'delegated powers in the interest of humankind'. " (Morgera, 2012) States, under these premises, are at the service of promoting intergenerational justice and well-being of individuals and certain groups within their own territory. To this end, global environmental law plays a crucial role in

[i]With more or less scepticism when talking of "Global Law," see for instance the works of Twining (2000, 2009); on the other side with a more enthusiastic approach, mooring into the lands of the clear emergence of some sort of global constitutional, see Walker (2012, 2008) and other of his most recent works that trace the theoretical framework of the foundations of global law with a very enthusiastic approach.

the implementation of the principle of common but differentiated responsibility under international law (Hey, 2010) that has been successfully promoted by the UNFCCC and the Kyoto Protocol.

Within this context, "sustainability" emerges as an "attempt" to become a global principle. The WTO has also taken important steps to protect the environment and to include the principle of sustainability into the dynamics of its work programme. Sustainable development is an objective of the WTO, as it was reflected in the Preamble of the Marrakesh Agreement Establishing the WTO, and the WTO case law has contributed to the legal reinforcement of the New Delhi principles of sustainable development although no case has been decided on the argument of "sustainable development" as such (Schurmans, 2015).

Specific types of global governance that shape the contours of future legal action are revealing the roads in which sustainability and "ecological" protection could successfully operate. This charming theoretical construction of a new global law, which arises out of the crisis of referring to the traditional institutional components of international law (sovereignty, territoriality, and the nation-state), is based on "the internationalization of human rights and the protection of the dignity of the individual that allows civil society to develop in the face of State hegemony" (Sanz Larruga, 2010). Following this line of reasoning it is important to refer to the works of Krisch and Kingsbury and Stewart about "The Emergence of Global Administrative Law" (Krisch et al., 2004); furthermore the work of Casini (Casini, 2006) is notable when referring to the particular importance of global administrative law in relation to issues of environmental law. The abovementioned works stress the idea of "environmental protection" as one of the areas that is rapidly developing and the development of global administrative law, underlying what is "[T]he vast increase in the reach and forms of transgovernmental regulation and administration designed to address the consequences of globalised interdependence" (Krisch et al., 2004). As far as the so-called "global administrative space" is concerned, evidence is given by a range of international and transnational regulatory schemes to which a variety of types of body are subject one obvious example related to the regulation of the environment which has arisen in the sphere of economic regulation (World Bank, OECD, WTO, etc.) in which long-range special regime regulatory structures are proliferating under the paradigm of sustainable development implementing the role of common but differentiated responsibility (Rajamani, 2006) (the Emissions Trading Scheme and the Kyoto Protocol Clean Development Mechanisms).

Moreover, a very significant idea, which is distinctive of global law (which distinguishes it from the classical regime of International law), is that those addressed by global regulatory regimes are increasingly the same as those under domestic law: individuals, companies, groups and even NGOs.[j] A good example should be the eventual situations whereby the purpose of global regulation aimed at the State is the protection of social and economic bodies, such as the environmental standards imposed by the World Bank for granting aid to developing countries. Kingsbury and Krisch and Stewart asserted that in such "multifaceted administrative space" where "states, individuals, firms, NGOs, and other groups or representatives of domestic and global social and economic interests who are affected by, or otherwise have a stake in, global regulatory governance, interact in complex ways" allowing "the relative autonomy and distinct character of this global administrative space, and its increasingly powerful decision-making bodies" (Krisch et al., 2004).

Local initiatives today are showing us the importance of joining efforts at the global level. A good example of this is given by the 11th Sustainable Development Goal on Urban Development. Cities around the world are engaged in facing the challenges of future urban development by integrating sustainability principles with all their local policies and creating worldwide networks (C40 cities, Covenant of Majors that was launched at European level in 2008 and extended to Eastern countries, Southern-Mediterranean, Sub-Saharan Africa, and in 2017, the launch of the Global Covenant of Mayors for Climate and Energy is expected).

Growing horizontal interconnectivity between policy domains is occurring, and finance, trade, culture, security, peace, and cooperation cannot be separated from sustainable development. Instead of clearly separated

[j]In this way, it becomes important to note that, in the environmental field, individuals and companies are direct subjects in the certification of CDMs (Clean Development Mechanisms), without any administrative intervention at all. Moreover, when certain international regulations have the aim to achieve specific changes in private behaviour, regulatory obligations are imposed on States and the way in which States monitor their private bodies is supervised, as for example, for the application of the Basel Convention on the Control of Transboundary Movement of Hazardous Waste and their disposal (Krisch et al., 2004).

areas of policy concerns and separate institutions to deal with them, there are now communities of different actors and layers that come together to create a global gathering place of multiple public and plural institutions (Stone, 2008). Thus, in a globalised world, important decisions related to environmental or natural resource issues may have more to do with corporate management and market outcomes than with state-level or other political factors creating "sustainable, equitable and democratic growth" (Stiglitz, 2003).

3.4 A European Vision of Sustainability: Normative Framework

Turning our focus on the EU context, the whole discourse becomes quite fascinating in terms of effective political objectives taking the lead on every policy area of the EU and in terms of tracing sustainability within its well-established constitutional framework that has evolved substantially from its original economic concerns. This contribution seeks to show how the EU vision of "sustainability" inherited from the global scenario that we have described operates in the current EU legal framework fitting in within traditional categories of normativity. The first premise starts by ascertaining how the EU is quite a uniquely successful experiment in regional environmental governance, especially after noting the lack of such legally binding operational governance at international and global levels. This is widely accepted considering how states have incrementally transferred sovereign rights to a supranational level (Bosselmann, 2008, p.187); of course, the obvious advantage of this is that member states coordinate their efforts, through binding (at different levels) legislation and centralised decision-making. Therefore to this end, the EU provides an exemplary model of governance for sustainability.

The EU immediately recognised the call for sustainability created by the international scenario, and it quickly added notions of sustainability into its constitutional framework. The two first steps of this adoption of sustainability norms into the EU as an incipient form of what they are today have been the fifth Environment Action Programme (EAP) and the Amsterdam Treaty on the European Union (TEU). The first EAP assumes the definition of sustainable development proposed by the Brundtland Commission with the objective to transform patterns of growth within the EU community in a manner that promotes sustainability (5th European Community Environment Programme: Towards Sustainability). The Treaty of Amsterdam amending the TEU embraced the internationally accepted definition of sustainable development proclaiming that "The Community shall have as its task promoting a harmonious, balanced and sustainable development of economic activities." The Treaty of Amsterdam also established a horizontal provision linked to sustainable development by introducing the principle of integration of environmental concerns into other EU policies with a view to promote sustainable development.[k] Therefore, within the EU Constitutional setting, sustainable development was initially deployed through the "integration principle"[l] that is now enshrined in article 11 of the Treaty on the Functioning of the European Union (TFEU), which provides environmental protection through the inclusion of environmental considerations at the earliest stage of the development process. But "sustainability" came prominently into life with the Lisbon Treaty.

The concept is currently enshrined in various treaty provisions, namely Article 3(3)–(5) TEU, Article 21 (2) (d)–(f), Article 11 TFEU, and Article 37 of the European Convention of Fundamental Rights (EUCFR). The definitions that are enshrined in the different instruments that represent the primary law of the EU and frame together the constitutional scenario of the EU do not refer exclusively to the development of economic activities. Article 3 (3) of the TEU reads as follows: "The Union...shall work for the sustainable development of Europe based on balanced economic growth and price stability, a highly competitive social market economy, aiming at full employment and social progress, and a high level of protection and improvement of the quality of the environment. It shall promote scientific

[k]Article 6 of the Treaty of Amsterdam Amending the TEU, the Treaties Establishing the European Communities and Related Acts (1997) OJ C340, reads as follows: "Environmental protection requirements must be integrated into the definition and implementation of the Community policies and activities referred to in Article 3, in particular with a view to promoting sustainable development."

[l]The integration principle finds its roots in the important *Danish Bottles* case. In that case, the ECJ held that a system requiring manufacturers and importers to market beer and soft drinks only in reusable containers (which had to be approved by a National Agency for the protection of the environment) was subject to the current Article 34 TFEU. The requirement implied a prohibition against the marketing of goods in containers other than ones that were returnable. The judgement is important as it enabled and facilitated the integration of environmental considerations into the market freedoms of the E.C. for ever!

and technological advance." Linking economic development, environment protection, and social justice has been a crucial objective.

However, it is important to understand that sustainable development, as well as environmental protection, cannot be separated from the internal market, regardless of the evolution of the concept towards a conciliatory nature[m] that is also reflected in thousands of secondary law instruments striving to create a consistent and separated area of *acquis communautaire* not yet openly recognised.[n] Still, the establishment of sustainable development as an objective, in its different forms by the constitutional framework of the EU means a step forward when considering nature preservation mandates regardless of its fragmented approach in secondary legislation (i.e. agriculture, technology, transport, hazardous waste, energy, renewable energy resources etc.).

Moreover, paragraph 5 of Article 3 of the TEU and Article 21 (2) (d) of the TEU, sustainable development also became one of the cornerstones of the EU external policy. The obligation to consider sustainable development in the EU international relations became remarkably important (Marín-Durán and Morgera, 2012). Furthermore, Article 11 of the TFEU

[m]The original appearance of such conciliatory nature was offered by a court decision by the ECJ; *First Corporate Shipping*, a case on development taking place in protected birds habitats, is testament to a conciliatory approach. Case C-371/98, *First Corporate Shipping*, (2000) E.C.R. I-9235, par. 54 (opinion AG Léger).

[n]It is possible to identify these aspects of the sustainability discourse in the EU through the initiatives of the EU's Environmental Action Programmes (EAP), the first of which was adopted in 1972 (First EAP (1972–1977), followed by others 1977–1982, 1982–1987, 1987–1992, 1992–2002, 2002–2012, and 2013–2020. The approach of the first EAP was to advocate the development of command and control initiatives designed to resolve specific pollution control problems, such as acid rain. This reflected the discourse of the period and demonstrated an approach that was unlikely to lead to an effective and comprehensive sustainable development strategy. However, it was evident by the time the EU adopted the Fifth EAP (1992–2000) that the approach had become closer to the "strong sustainability" position. The changing nature of the commitment to a strategy for sustainable development was shown in the use, for the first time, of a specific title for the Action Programme "Towards Sustainability." The initiatives proposed in the Fifth EAP concentrated on measures targeting specific sectoral policies and priorities in order to achieve the sustainable development objectives. The discourse had shifted from its origins as action on pollution control to a model based on ecological modernisation as the dominant ideology. See Chapters 2 and 13 of on this point. The seventh EP "Living well within the limits of our planet" shows an even further commitment to sustainability approaching all the aspects that constitute the challenges of our present time, fostering a holistic vision of sustainability that goes beyond an anthropocentric approach to build on a more eco-centric approach, that brings about social justice and good governance.

re-proposes the integration principle; it states that: "Environmental protection requirements must be integrated into the definition and implementation of the Union Policies and activities, in particular with a view to promoting sustainable development," which is not so different from this formulation, Article 37 of the EUCFR reads: "A high level of environmental protection and the improvement of the quality of the environment must be integrated into the policies of the Union and ensured in accordance with the principle of sustainable development."

The fact that sustainable development is condensed in these different provisions placed at the top of the hierarchical pyramid of the EU legal order does not mean that its legal status is not minimised by disputes. As Sadeleer puts it, sustainable development has a "status dogged by controversies" (Sadeleer, 2015), thus causing some problems perhaps regarding conceptual and status interpretation; these conflicts might lead to a reluctance by the European Union Institutions (specially the CJEU) to develop and stick to a firm interpretation of sustainability[o] as a hermeneutic instrument for case-law resolution. There are no cases decided on the basis of sustainable development, despite the existence of recent crucial cases wherein the court has tackled its definition in certain aspects of the EU Union cooperation policy after Lisbon[p] and on the interpretative features of its sectorial legislation.[q]

Could sustainability really become a normative concept within EU law? Authors like Sadeleer have no problems affirming that "combined with the requirements of integration, a high level of protection, and the different principles of environmental law (prevention, precaution, polluter-pays, etc.) sustainable development had become a normative concept" (Sadeleer, 2015); it is undoubtedly a binding constitutional objective. Therefore, rather than being sceptical about the possibility of sustainability to represent a primary paradigm of the numerous EU policies, we should examine its quick evolution and the crucial role of the EU trying to coordinate its

[o]See the interpretation given by AG Léger to sustainable development in its opinion in *First Corporate Shipping*, a case on development taking place in protected birds habitats, is testament to a conciliatory approach. Case C-371/98, *First Corporate Shippping*, (2000) E.C.R. I-9235, par. 54 (opinion AG Léger).

[p]C-377/12, *European Commission v Council of European Union* (Phlippines PCFA) EU:C:2014:1903, Judgement of the Court Grand Chamber.

[q]C-461/13, *Bund v Germany*, ECLI:EU:C:2014:2324 (opinion AG Jääskinen). See further, Paloniitty (2016).

member states to fight for ecological problems, energy supply, technological development, and local initiatives for good governance following the action plan of the "Covenant of Majors." Another aspect that should not be forgotten, which is related to the discourse of a low-carbon "green" economy, is focusing on job creation. It became a nodal point in the political discourse[r] and it received support from the European Parliament, the European Commission and some of the EU member states, particularly Germany and the UK. To achieve the low-carbon economy or "green economy," three things are necessary: environment protection, job creation and secure energy supply as "[T]he entire industrial infrastructure, based on the back of fossil fuels, is ageing and in disrepair." (Rifkin, 2011). Rifkin argued that putting the environment at the centre of the economy would help to achieve the sustainable use of natural resources. Thus, there would be environmental protection, energy security and economic benefits from making a transition to a "green economy." Although the discourse on the green economy has gained support (i.e. the Italian government has enacted a law following this conceptual trend), it has not replaced sustainable development as the hegemonic discourse; rather, it appears to offer support for the discourse identifying an agenda of action for sustainable development. Moreover, the scientific and political discourses have turned towards more social policy concerns and market- and industry-oriented approaches focusing on the transition to a low carbon economy and natural resources and their sustainable lifecycle as well as how to incorporate corporate social responsibility improve ethical and environmentally friendly industrial performance. So, let's say that at the EU level, the *acquis communautaire* on sustainability, once we have established its constitutional and foundational basis, there are four main challenges at the forefront: first, the implementation of an ambitious agenda to combat climate change and create the basis for an energy transition and low-carbon economy; second, the circular economy package should be implemented and somehow culturally embedded and promoted in member state production and consumption habits; moreover, every social policy is, or rather should be, related to the European Sustainability Model that European Institutional bodies should support and promote.

---

[r]Based on the writings of Rifkin (2011) and NEP (2011).

EU legislation in the field of "sustainability" can only be understood in light of the parallel effort of the international agreements for the determination of concepts, categories and basic principles for the development of a principle of intergenerational justice. Among these legal categories, "sustainability" should be remarked, without giving up a dynamic speech challenging its substance. In fact, regardless of the hegemonic character of "sustainability," the global legal order in which different levels of governance interact modifying and developing the protection keys aiming at compliance with a supranational criteria have focused their attention on climate change talks. As a consequence, the challenges that sustainable development and sustainability have posed to public policy have often been dwarfed in economic speeches. Currently, the stress is placed on the ambitious UN Global Agenda and the SDG that the EU has incorporated to its political mandate.

3.5 EU Sustainability Model: Work in Progress with Many Challenges

The incorporation of the "sustainability" concept into the EU treaties still reflects a great ambivalence, but the main challenge for the EU is to overcome the fact that its legal significance is not that of a legal principle. Thus, sustainable development is not legally justifiable for it does not lead (yet) to a specific obligation, which would be enforceable before the CJEU. In fact, the EU legal commitments to sustainable development have so far been weaker than the political ones. Nevertheless, specific legal packages and directives in different policy areas where EU competence prevails over the national one, have determined some changes in crucial areas such as energy and climate change, waste, water, food, and public procurement. Some coordinated actions at the local level in which European institutional support plays an important role have also been developed. A key factor for those subtle developments for a strong European sustainability model is determined by the necessary cooperation to enable further and stronger innovation and cooperation (under sustainable premises) in energy transition, climate change policy, human rights and access to justice issues, waste management, and sustainable consumption and production.

Still, the case study of the EU's energy policy highlights the manner in which the comprehensive sustainable development paradigm has been somehow vanished by a narrow definition of sustainability; this is because it has been perceived too often as the dimension of the energy policy focusing on the reduction of Greenhouse Gas (GHG) emissions to combat climate change. Nonetheless the energy policy-making process at EU level, and the significant changes introduced at the constitutional level in this field,[s] have fostered another *acquis* of communitarian origin fighting for a "sustainable European making" EU Energy policy has focused too much on member states interest on decarbonisation without attaining to common policies, nor transferring good practices (Sánchez Galera, 2016).

Despite the many deficiencies, the EU can export a governance model shaped by the sustainability paradigm that was originally launched by the global scenario. Looking at concrete initiatives at the European level (on energy, trade, agriculture development, spatial planning, etc.), or the national level (i.e., French Energy Transition Law of July 2015, Finnish full decarbonisation initiative, German phasing out nuclear power) the EU is on the good track towards the improvement of the social and environmental well-being within member states regardless of its original functioning on market-based documents. The most important legal instruments of this model that will shape the future *acquis communitaire* related to a sustainable policy are represented by the most recent circular economy package and the successful European Climate and Energy system. The EU is the world leader in climate policy, having already taken important steps for a sustainable, low carbon future. The most recent Energy Roadmap 2050 calls nearly for complete elimination (80–95 % reduction) of GHG by mid-century. The "Clean Energy for all Europeans' Package" has been approved, and the EU is in the process of updating its energy policy framework in a way that will facilitate clean energy transition. This is a very significant, although, complex step for the creation of the Energy Union

[s]Firstly, note that the first place the so-called Cardiff process (regardless of its unsuccessful development, launched in 1998, concerns the integration of the environment into sectorial policies for sustainable development. Despite several summits discussing the Cardiff Process with a view to develop a comprehensive and integrated strategy for sustainability, it remained unclear how and to what extent the Cardiff process can be linked to the agenda of sustainability. Furthermore, the Treaty of Lisbon introduces a chapter about energy. Energy Policy can be categorised as "brand new" into the sovereign domain of the EU, suggesting in its wording "sustainability" with a conciliatory nature towards some ecological concerns, encapsulated in Article 194 and Article 195 of the TFEU.

and delivering on the EU's Paris agreement commitments. An important and innovative element of the new policy framework is the "regulatory" certainty, in particular through the introduction of the first national energy and climate plans, encouraging essential investments to take place in this sector. Another essential element is the full integration and empowerment of European consumers to become fully active players in the energy transition. It also fixes two new targets for the EU for 2030: a binding renewable energy target of at least 32% and an energy efficiency target of at least 32.5%, with a possible upward revision in 2023. For the electricity market, it confirms the 2030 interconnection target of 15%, following on from the 10% target for 2020.

The previous 20/20/20 by 2020 had three key objectives: a 20% reduction in the EU GHG emissions from 1990 levels; raising the share of EU energy consumption produced from renewable resources to 20%; and a 20% improvement in the EU's energy efficiency. At the level of EU as a whole, there is EU Emissions Trading System (EU ETS) working on the "cap and trade" principle and currently, being in its third phase (2013–2020), a single, EU-wide cap on emissions applies in place of the previous system of national caps. The Energy Union project seeks to further integrate the internal market and to develop a more common approach to secure energy supply by creating a genuine solidarity basis, which underlies the climate change global dimension that integrates our energy policy and shapes European Energy Governance on the basis of solidarity and cooperation. Unfortunately, the interaction of unexpected political, social or economic components could create distortions that the legal framework might have not foreseen, but the new ambitious targets will stimulate Europe's industrial competitiveness, reduce energy bills, create competitive jobs and tackle energy deficiency while improving air quality.

Because of how domestic politics are currently evolving throughout Europe, important issues, such as migration and liberal democracy appear to drift from consensus, and the European integration dream is hampered by the heterogeneity of political views and withdrawal momentum. It remains to be seen whether the "sustainability model" based on areas that have traditionally relied on broad consensus (i.e. climate change, agriculture, food, waste, and water management) does not touch issues too much related to "sovereignty" and manages to coordinate institutionally a smart programme to innovate and help member states to accelerate their societies into ecological constraints and awareness.

The regulatory instruments and the amount of binding legislation, communication, recommendations, etc., from the EU Commission have rendered the whole of environmental law, climate change, circular economy package and development and cooperation policy, extremely complicated and tremendously ambitious, but the European coordinated action has so far brought success and leadership to Europe and European member states singularly to face the major social and economic transitions within sustainability constraints.

The Circular Economy Action Plan was adopted by the European Commission in December 2015, which is an ambitious new Circular Economy Package launched to stimulate Europe's transition to a circular economy that will boost global competitiveness, foster sustainable economic growth and generate new jobs.[t] The European Commission has already presented its report in which all 54 actions foreseen under the Circular Economy Action Plan have now been delivered or are being implemented, thus paving the way towards a climate-neutral, competitive circular economy where pressure on natural and freshwater resources, as well as ecosystems, is minimised.[u] Through the Circular Economy Action Plan the Commission has mainstreamed circular principles into plastic production and consumption, (…) water management, food systems and the management of specific waste streams. The revised waste legislation,[v] so unique and definitely a blueprint for the EU sustainability model, requires that by 2030, 70% of all packaging waste and, by 2035, 65% of all municipal waste should be recycled, while reducing landfilling of municipal waste to 10%. A 5-year time extension is granted to several countries.[w]

---

[t]Europa.eu/rapid/press-release_IP-15-6203_en.pdf

[u]See Report from the Commission to the European Parliament, the Council, the European Economic and Social Committee and the Committee of the Regions on the implementation of the Circular Economy Action Plan, Brussels, 4.3.2019, COM (2019) 190 final.

[v]OJ, 14.62018, L 150, p 93, 100, 109, 141 Directive 2008/98/EC on waste, Directive 1999/31/EC on the landfill of waste, Directive 94/62/EC on packaging and packaging waste, Directive 2000/53/EC on end-of life vehicles, Directive 2006/66/EC on batteries and accumulators and waste batteries and accumulators, Directive 2012/19/EU on waste electrical and electronic equipment (WEEE).

[w]Greece, Croatia, Cyprus, Latvia, Lithuania, Hungary, Malta, Romania, Slovakia and Bulgaria.

The 2030 Agenda for Sustainable Development is rooted into the international cooperation and development policy of the EU and there is a strong and renewed emphasis on the EU's global strategy. The SDGs will be a cross-cutting dimension of the implementation of the EU's global strategy. "The European Consensus on Development" (*European Consensus on Development: Our World, Our Dignity, Our Future*, Official Journal of the EU, C 210, 30 June 2017) is a blueprint aligning the Union's development policy with the 2030 Agenda for Sustainable Development in order to strengthen the EU's response to the 2030 Agenda to improve coordinated implementation and enforcement as well as to improve EU impact. There are many cross-cutting topics as part of the development and cooperation agenda in which the EU model could lead a global action, that is, youth, gender equality, mobility and migration, sustainable energy and climate change, investment and trade, good governance, democracy, the rule of law and human rights, innovative engagement with more advanced developing countries, and mobilising and using domestic resources.

A clearly successful new paradigm of sustainable development that is building a new body of *acquis communitaire* in all European sectorial policies is here to stay despite its complexities and its abundant and prolific set of normative components. Furthermore, it will impact data documents, performance-related reports as well as the proliferation of bottom-up approaches that are currently gaining momentum but for limitations of space will not be discussed here.

Conclusions

Sustainability, undoubtedly, is a new paradigm of action at every level of governance worldwide. Despite the difficulties involved in the lack of consensus on the meaning of "sustainable development" at the global level, considering that the debate on its conceptual substance is far from resolved, at the very least this concept calls for an integrated consideration of social, economic, and environmental concerns. The multidimensional nature of the concept and its materialisation as a political objective is today fulfilled with a growing array of values, with sustainable development being placed among the peace and security mandatory objectives of the UN despite its clear deficiency on enforcing mechanisms.

At the regional level, the EU's Sustainable Development Strategy provides an appropriate framework to analyse the changing nature of the discourse in the developed world. The meaning and legal force of sustainability at the European level constitutionally enforced (although it has a long way to go in order to foster genuinely sustainable development) has become a unique model of effective governance; such governance could be used to implement sustainable development through policies and binding instruments that could change the course of events even at the global level through successful climate change, production, and waste and consumption policy that go beyond market integration interests. Still, despite the centrality of climate change concerns in the "rhetoric" of the European Commission an effective integration of environmental goals merged with the sustainability criteria in fields such as the energy policy field is not so easy to achieve. Nevertheless, a clearly successful new paradigm of sustainable development that will build a new body of *acquis communitaire* within ecological constraints governs all European sectoral policies and is here to stay. The CJEU could play in the future an important role on litigation matters concerning compliance with binding sustainability norms, or related prescriptions that are still missing today.

Within the EU context, the TFEU and TEU give us guidelines for a strong interpretation of a conceptual and functional model of sustainable development, as it was given at the time by the Brundtland Report. Still, regardless of its successful settlement with an hegemonic position in the global arena with regard to the conciliation of economic demands with sustainable growth and development attached to practices of good governance, there seems to be no clear recognition of its binding nature, neither its status as a recognised and undiscussed principle of International law despite the consensus reached on the Sustainable Development Agenda adopted by the UN concluding a negotiating process that has spanned more than two years and has featured unprecedented participation of civil society. It is at the EU governance level that we can recognise a model of an effective enforcement system aiming at "sustainable development." The integration and balancing exercise called by sustainable development and the global agenda, cannot sacrifice by any means ecological integrity in order to succeed, especially in a world (still) strongly conceptualised in economic terms. Its proactive nature, and dynamic and contextual normative existence has led recent scholars to emphasise its social dimension as an open-wide prerequisite of development accompanied by dimensions of peace, security and even culture (Rodrigo, 2015, p. 39) and has determined its prominent existence shadowing the concept of "green economy" and reaching better acceptance within normative and political discourses.

Briefly, by assessing primary and secondary law, it is possible to see how EU's normative framework on sustainable development is clearly embedded in its constitutional texts. On the basis of such constitutional mandate and concrete consensus at the political level on cross-cutting topics of environmental law nature, secondary legislation becomes more and more complex, and although energy and climate policy take the lead to define the content of European sustainability policy, there is space for production and consumption, waste, and social issues. Today, we can confidently assert that Europe has built an ambitious body of *acquis communautaire* in sustainability.

Thus far, the evolution of EU environmental law and policy since the first Environmental Framework Programme has offered the evidence of how the European project has remarkably expanded from the specific economic sphere of market integration to address new social and global challenges. The Europe 2020 Strategy and its leading initiative on a "Resource Efficient Europe," the circular economy package, the "clean energy for all" package and its most recent consensus on development reflect this new approach and sets the path for a transformation of European economy from resource intensive to resource efficient in line with worldwide efforts to achieve the transformation for a green economy and the eradication of poverty. The newly approved clean energy package strives for long-term GHG reduction creating a strong basis for solidarity, cooperation and investment. The new rules put consumers at the heart of the transition and this empowerment of the civil society component also sheds new light on the EU governance model.

A multipolar world led by a complex web of relations has emerged. Incidental to this there are mostly four types of actors leading global and regional governance mechanisms with statehood properties (i.e. global institutions, regional organisations, states and subnational regional entities) together with non-state actors such as NGOs or transnational policy networks. This means a new conceptualisation of governance, citizenship and dialogue in international/global relations therefore requires a multiplicity of citizenships as a political-legal status, a recognition of diverse and multiple identities (i.e. multiculturalism), and citizens' participation in all levels of sovereignty (i.e. transnationalism). In addition to this, the growing awareness of the interconnection of the world and the transnational effects of internal policy approaches, not only on trade but also in other policy areas, is expected to lead further international cooperation for the attainment of the UN SDG that are the universal expression of a world lead by the sustainability paradigm.

References

Arias-Maldonado, M. (2015) *Environment and Society: Socionatural Relations in the Anthropocene*. Springer.

Barnes, P. and Hoerber, T. (eds.) (2013) *Sustainable Development and Governance in Europe*. Routledge, UK.

Barral, V. (2012) Sustainable development in international law: Nature and operation of an evolutive legal norm. *European Journal of International Law*, 23(2), 377–400.

Beck, U. (2007) *What is Globalisation?* Polity Press, Cambridge, UK.

Birnie, P. W. and Boyle, A.E. (1992) *International Law and the Environment*. Oxford, UK.

Bosselmann, K. (2008) *The Principle of Sustainability: Transforming Law and Governance*. Ashgate, Farhan, UK.

Bosselmann, K. (2010) Sustainability and the courts: A journey yet to begin? *Journal of Court Innovation*, 3(1), 338.

Caradonna, J.L. (2014) *Sustainability. A History*. Oxford University Press, New York.

Carlowitz, H.C. von (1713) *Sylvicultura oeconomica, oder haußwirthliche Nachricht und Naturmäßige Anweisung zur wilden Baum-Zucht*. TU Bergakademie Freiberg und Akademische Buchhandlung: Leipzig, repr. Freiberg.

Casini, L. (2006) *Diritto Amministrativo Globale in Dizionario di Diritto Pubblico*, S. Cassese (Dir.), Giuffrè, Milano.

Cordonier Segger, M.C. and Khalfan, A. (2004) *Sustainable Development Law: Principles, Practices and Prospects*. Oxford University Press, UK.

Esty, C. and Ivanova, M. (2004) *Globalisation and Environmental Protection: A Global Governance Perspective,* available at: www.Yale.edu/environcenter/yaleglobal.yale.edu

Evelyn, J. (1664) *Sylva, or a Discourse of Forest-Trees and the Propagation of Timber in His Majesties Dominions.* London, Printed by J. Martyn and J. Allestry.

Grober, U. (2007) *Deep Roots. A Conceptual History of 'Sustainable Development' (Nachhaltigkeit).* Wissenschaftszentrum Berlin für Sozialforschung (WZB), Berlin.

Hey, E. (2010) Global environmental law and global institutions: A system lacking "Good Process," in Pierik, R. and Werner, W. (eds.), Cosmopolitanism in Context: Perspectives from International Law and Political Theory, Cambridge 45 and 50.

Krisch, N., Kingsbury, B. and Stewart, R.B. (2004) The Emergence of Global Administrative Law, *IILJ Working Paper.*

Lowe, V. (1999) Sustainable development and unsustainable arguments, in Boyle, A. and Freestone, D. (eds.), *International Law and Sustainable Development: Past Achievements and Future Challenges.* Oxford, Oxford University Press.

Marín-Durán, G. and Morgera, E. (2012) *Environmental Integration in the EU's External Relations.* Hart Publishing, UK.

McAfee, K. (1999) Selling nature to save it? Biodiversity and the rise of green developmentalism, *Environment and Planning,* 17, 2.

Morgera, E. (2012) Bilateralism at the Service of Community Interests? Non-judicial Enforcement of Global Public Goods in the Context of Global Environmental Law. *European Journal of International Law,* 23(3), 743–767.

Paloniitty, T. (2016) The Weser case: Case 461/13 bun v Germany, *Journal of Environmental Law,* 28(1), 151–158.

Percival, L.J. and Piermattei, W. (eds.) (2014) *Global Environmental Law at the Crossroads.* Edward Elgar pub.

Rajamani, L. (2006) Differential Treatment in International Environmental Law, Oxford University Press, UK.

Rifkin, J. (2011) *The Third Industrial Revolution: How Lateral Power is Transforming Energy, the Economy and the World.* Palgrave Macmillan, and NEP: Basingstoke and New York.

Robinson, J. (2004) Squaring the Circle? Some thoughts on the idea of sustainable development. *Ecological Economics,* 48, 370.

Rodrigo, A.J. (2015) *El desafío del desarrollo sostenible. Los principios de Derecho internacional relativos al desarrollo sostenible.* Centro de Estudios Internacionales, Marcial Pons, Madrid.

Sachs, J.D. (2015) *The Age of Sustainable Development.* Columbia University Press, New York.

Sadeleer, N. (2015) Sustainable Development in EU Law: Still a Long Way to Go. *Jindal Global Law Review*. Special Issue on Environmental Law and Governance (2015) 6(1), 39–60.

Sánchez Galera, M.D. (2016) *La integración de las políticas energéticas y ambientales en la Unión Europea: Paradojas y Contradicciones a la luz del paradigma de la sostenibilidad*, Revista General de Derecho Administrativo de Iustel, 43, 1–28.

Sand, P. (2006) Global environmental change and the nation state, in Winter, G. (ed.), *Multilevel Governance of Global Environmental Change*. Cambridge University Press, UK.

Sands, P. (2003) *Principles of International Environmental Law*, 2nd edition. Cambridge University Press, UK.

Sanz Larruga, J. (2010) Environmental law and its relationship with global administrative Law, in Robalino-Orellana, J. and Rodríguez-Arana Muñoz, J. (eds.), *Global Administrative Law. Towards a Lex Administrativa*. Cameron May CMP Publishing, London.

Schurmans, M. (April 2015) Sustainable development is emerging as a core tenet of WTO case law. To what extent has it helped enshrine this as a legal concept? *European Energy and Environmental Law Review*, 28, 28-34.

Stewart, R.B. (2014) Addressing problems of disregard in global regulatory governance: Accountability, participation and responsiveness, *IILJ Working Paper* 2014/2 (Global Administrative Law Series).

Stiglitz, J.E. (2003) *Globalization and its Discontents*. Norton Paper Back, New York.

Stone, D. (2008) Global public policy, transnational policy communities and their networks. *Journal of Policy Sciences*, 36(10), 19–38.

Twining, W. (2000) *Globalisation and Legal Scholarship*. Cambridge University Press, UK.

Twining, W. (2009) *General Jurisprudence: Understanding Law from a Global Perspective*. Cambridge University Press, UK.

Walker, N. (2012) Constitutional pluralism in global context, in Avbelj, M. and Komárek, J. (eds.), *Constitutional Pluralism in the European Union and Beyond*. Hart: Oxford.

The World Summit on Sustainable Development (2002) Johannesburg, South Africa, 26th August-4th September, Document retrieved from UNdocuments.net

Yang, T. and Percival, R.V. (2009) The emergence of global environmental law. *Ecology Law Quarterly*, 36, 615.

Charters, Resolutions, Reports, Treaties

Aalborg Charter of European Cities and Towns Towards Sustainability (1994).

EU's Environmental Action Programmes (EAP), the first of which was adopted in 1972 (First Environmental Action Programme (1972–1977), followed by others in 1977–1982, 1982–1987, 1987–1992, 1992–2002, 2002–2012, and 2013–2020.

European Consensus on Development - Our world, our dignity, our future, 8 June, 2017, accessible at ec.europa.eu

"Fifth European Community Environment Programme: Towards Sustainability" available at: https://eur-lex.europa.eu/legal-content/EN/TXT/HTML/?uri=LEGISSUM:l28062&from=EN, accessed December 2015.

General Assembly Resolution S-19/2, 28th June 1997.

Johannesburg Declaration on Sustainable Development (Johannesburg Declaration) Johannesburg, 4 September 2002, UN Doc. A/Conf. 199/20 (2002).

New Delhi Declaration on the Principles of International Law Related to Sustainable Development (London, 2002; ILA resolution 3/2002).

Report from the Commission to the European Parliament, the Council, the European Economic and Social Committee and the Committee of the Regions on the implementation of the Circular Economy Action Plan, Brussels, 4.3.2019 COM (2019) 190 final.

Rio Declaration on Environment and Development (1992) A/CONF. 151/26 (vol. I).

Treaties Establishing the European Communities and Related Acts (1997) OJ C340

United Nations (2014) *Prototype Global Sustainable Development Report* New York: United Nations Department of Economic and Social Affairs, Division for Sustainable Development. https://sustainabledevelopment.un.org/globalsdreport/

World Commission on Environment and Development (UNWCED) (1987) Our Common Future, 'Brundtland Report' (Oxford and New York, Oxford University Press).

Case Law

Case C-371/98, First Corporate Shipping (2000) E.C.R. I-9235, par. 54 (opinion AG Léger).

C-461/13, Bund v. *Germany, ECLI:EU:C:2014:2324 (opinion AG Jääskinen).*

C-377/12, European Commission v. *Council of European Union (Phlippines PCFA) EU:C:2014:1903, Judgement of the Court Grand Chamber.*

Gabčikovo-Nagymaros Dam Case (1997) ICJ Reports 7, para. 140.

Iron Rhine Railway Railway Arbitration (2005) PCA, paras. 58–59.

Pulp Mills Case (Provisional Measures) (Argentina v. *Uruguay) (2006) ICJ Reports, para. 80.*

Pulp Mills Case (Merits) (2010) ICJ Reports, para. 177.

CHAPTER 4

Climate Change and Development Cooperation in the European Union

Kattya Cascante Hernández

(...) In a moment when the crucial role of the United Nations' system, the importance of development cooperation, or the reality of climate change are put into question, the Global Strategy has been a reminder of the European Union's strategic interest in a cooperative world order.

–Federica Mogherini in EEAS (2017)

A good society is not only an economically prosperous society, but it has to be also socially inclusive, environmentally sustainable and governed well.

–Sachs (2015)

Introduction

This chapter analyses how the commitments adopted in the Agenda 2030[a] have impacted European cooperation and the development cooperation policies of the European Union (EU).

[a]The 2030 Agenda for Sustainable Development adopted at the United Nations Sustainable Development Summit on 25 September 2015. This Agenda is a plan of action for people, planet and prosperity. It also seeks to strengthen universal peace in larger freedom. "We recognise that eradicating poverty in all its forms and dimensions, including extreme poverty, is the greatest global challenge and an indispensable requirement for sustainable development" (UN, 2015a). All countries and all stakeholders, acting in collaborative partnership, will implement this plan. The 17 Sustainable Development Goals and 169 targets demonstrate the scale and ambition of this new universal Agenda with the three dimensions of sustainable development: the economic, social and environmental.

Article 21 of the Treaty of the European Union (EU) (2012)[b] establishes the principles governing its external action and its relations with third countries as well as with other international organisations, whether regional or global, which share those principles. As such, the EU commits itself to seek and adopt multilateral solutions to collective problems, always within the framework of the EU. More precisely, in point d) of this article the EU commits to "foster the sustainable economic, social and environmental development of developing countries, with the primary aim of eradicating poverty" (EU, 2012).[c] Accordingly, the EU endorses the definition of sustainable development approved in the World Development Summit of 1987[d]; a definition that was subsequently developed in the 1992 Earth Summit Celebrate in Rio,[e] the 2002 Johannesburg Summit[f] and the 2012

[b]The Union's action on the international scene shall be guided by the principles, which have inspired its own creation, development and enlargement, and which it seeks to advance in the wider world: democracy, the rule of law, the universality and indivisibility of human rights and fundamental freedoms, respect for human dignity, the principles of equality and solidarity and respect for the principles of the United Nations Charter and international law (Art 21, EU, 2012).

[c]Consolidated version of the Treaty on the Functioning of the European Union, *OJ C 326, 26.10.2012, 0001 - 0390.*

[d]"Our Common Future" also called the Brundtland Report, publication released in 1987 by the World Commission on Environment and Development (WCED), which introduced the concept of sustainable development and described how it could be achieved. Sponsored by the United Nations (UN) and chaired by Norwegian Prime Minister Gro Harlem Brundtland, the WCED explored the causes of environmental degradation, attempted to understand the interconnections between social equity, economic growth, and environmental problems, and developed policy solutions that integrated all three áreas (UN, 1987).

[e]The Rio Declaration on Environment and Development (UN (1992) reaffirming the Declaration of the United Nations Conference on Human Environment, adopted at Stockholm on 16 June 1972. Agenda 21 was a global consensus document. It is a vast work programme for the 21st century, approved by consensus among the world leaders in Rio, representing over 98% of the world's population. A comprehensive blueprint for a global partnership, Agenda 21 strives to reconcile the twin requirements of a high-quality environment and a healthy economy for all people of the world, while identifying key areas of responsibility as well as offering preliminary cost estimates for success (UN, 1992).

[f]The World Summit on Sustainable Development (WSSD), or the ONG Earth Summit of 2002 (also informally nicknamed "Rio+10"), took place in Johannesburg, instead of fostering the adoption of new agreements between governments, the Earth Summit was organised mainly around almost 300 "partnership initiatives." These were to be the key means to achieve the Millennium Development Goals by means of the Implementation Plan of the Summit, which focused on achieving goals of sustainable development by developing more effective, democratic and accountable international and multilateral institutions. (UN, 2002).

Rio 20[g] and thereafter turned into the central element of the Sustainable Development Agenda (SDGs)[h] and the Paris Agreement on climate change, which addressed the need to limit the rise of global temperatures.[i]

Sustainable development has been defined in many ways, but the most frequently quoted definition is from *Our Common Future*, also known as the Brundtland Report: "Sustainable development is development that meets the needs of the present without compromising the ability of future generations to meet their own needs" (UN, 1987). With this definition in mind, this chapter examines the way in which the EU's Development Cooperation Policy is implementing Agenda 2030. First, it revises the commitments adopted by the EU in relation to the Development Aid Effectiveness and Financing principle to promote sustainable development. Second, it analyses how the EU ensures coherence among the commitments it has adopted in different international fora (regarding, e.g., trade, climate change, or environment, social and financial issues); whether the EU distributes fairly its financial resources among them, and, finally, whether EU policies strengthen multilateral institutions.

[g]The Rio + 20 UN (1992) resolved to take concrete measures in order to accelerate the implementation of sustainable development commitments, to renew the commitment to sustainable development, assess the progress achieved to the date and the remaining gaps in the implementation of the outcomes of the major summits as well as address the new and emerging challenges. The Rio + 20 focused on two topics: a green economy in the context of sustainable development together with poverty eradication, and the institutional framework for sustainable development (UN, 2012).

[h]On 1 January 2016, the 17 Sustainable Development Goals (SDGs) of the 2030 Agenda for Sustainable Development—adopted by world leaders in September 2015 at a historic UN Summit—officially came into force. Over the next 15 years, with these new Goals that apply universally to all, countries will mobilise efforts to end all forms of poverty, fight inequalities and tackle climate change, while ensuring that no one is left behind (UN, 2015a).

[i]For the first time, the Paris Agreement brings all nations under a common cause: to undertake ambitious efforts to combat climate change and adapt to its effects, with enhanced support to assist developing countries to do so. The central aim of the Paris Agreement is to strengthen the global response to the threat of climate change by keeping a global temperature rise this century well below 2°C above pre-industrial levels and to pursue efforts to limit the temperature increase even further to 1.5°C. Additionally, the agreement aims to strengthen the ability of countries to deal with the impacts of climate change (UNFCCC, 2016).

4.1 The Evolution of the Goals of the International Development Cooperation Agenda

Though Development Aid is a long-standing policy domain that finds its origins at the end of IIWW,[j] this chapter only addresses its evolution in the present century, starting in the year 2000, when, for the first time, a Universal Agenda to face the challenges of development was defined.

This section examines the development agendas that have been successively adopted since then. First, it examines the two conferences dedicated to raising the funds required to achieve the 2000–2015 Millenium Goals (held in Monterrey in 2002 and in Doha in 2008). Second, it examines the new agenda focusing on effectiveness, which resulted from the insufficient funding raised by such conferences and the consequent donor's fatigue; and agenda that was articulated in four main conferences: Rome (2003), Paris (2005), Accra (2008) and Busán (2011), together with the European Consensus reached in 2006.

The United Nations (UN) Millennium Development Goals (MDGs) were adopted by the world's leaders at the UN Summit of September 2000. They constitute an internationally agreed-upon framework of 8 goals and 18 targets, which are time-bound and quantified, addressing extreme poverty, with a deadline of 2015. At the same time, the Poverty Reduction Strategies (PRSs) were introduced by the World Bank (WB) and the International Monetary Fund (IMF) in 1999,[k] as a concrete action plan

[j]The Marshall Plan (officially the European Recovery Programme, ERP) was an American initiative to aid Western Europe in which the United States gave over $12 billion (nearly $100 billion in 2016 US dollars) in economic assistance to help rebuild Western European economies after the end of World War II. The goals of the United States were to rebuild war-torn regions, remove trade barriers, modernise industry, improve European prosperity, and prevent the spread of Communism (Ryland and Samuel, 2018).

[k]In September 1999, the IMF established the Poverty Reduction and Growth Facility (PRGF) to make the objectives of poverty reduction and growth more central to lending operations in its poorest member countries. PRGF-supported programmes are framed around comprehensive, country-owned Poverty Reduction Strategy Papers (PRSPs). PRSPs are prepared by governments with the active participation of civil society and other development partners. PRSPs are then considered by the Executive Boards of the IMF and World Bank as the basis for concessional lending from each institution and debt relief under the joint Heavily Indebted Poor Countries (HIPC) Initiative. The targets and policy conditions in a PRGF-supported programme are drawn from the country's PRSP.

for poverty reduction, formulated by developing countries themselves. The principles that guide the formulation of PRS are country ownership and stakeholder partnership. Indeed, such strategies are to be formulated by poor countries applying for the Enhanced Heavily Indebted Poor Country (HIPC) Initiative,[1] together with the WB, the International Development Association (IDA), the IMF and the Poverty Reduction and Growth Facility (PRGF) financial 147 heads of state gathered at the UN Millennium Summit in 2000. They adopted the MDGs to address extreme poverty in its many dimensions—income poverty, hunger, disease, lack of adequate shelter and exclusion—while promoting education, gender equality, and environmental sustainability, setting quantitative targets in relation to each. This MDGs Agenda was the result of a two-year consultation process covering issues such as poverty eradication, environmental protection, human rights and the protection of the most vulnerable. All 191 UN member states at that time as well as 22 international organisations committed to helping achieve the MDGs by 2015 (Table 4.1).

Table 4.1 The Millennium Development Goals (2000–2015).

1. To eradicate extreme poverty and hunger
2. To achieve universal primary education
3. To promote gender equality and empower women
4. To reduce child mortality
5. To improve maternal health
6. To combat HIV/AIDS, malaria, and other diseases
7. To ensure environmental sustainability
8. To develop a global partnership for development

Source: UN (2000).

The MDGs emphasised three areas: human capital, infrastructure and human rights (social, economic and political), with the intent of increasing living standards. First, the human capital objectives include nutrition, education and healthcare (including child mortality, tuberculosis, malaria,

[1]The World Bank, the International Monetary Fund (IMF) and other multilateral, bilateral and commercial creditors began the Heavily Indebted Poor Country (HIPC) Initiative in 1996. The structured programme was designed to ensure that the poorest countries in the world were not overwhelmed by unmanageable or unsustainable debt burdens. It reduces the debt of countries meeting strict criteria (World Bank, 2018).

HIV/AIDS and reproductive health). Second, the objectives regarding infrastructure include access to safe drinking water, energy and modern information/communication technology; increased farm outputs using sustainable practices; transportation and environment. Finally, the human rights objectives include empowering women, reducing violence, increasing political voice, ensuring equal access to public services and increasing security of property rights. The goals were intended to increase an individual's human capabilities and "advance the means to a productive life." Objective 7 addresses environmental sustainability through four goals: first, to integrate the principles of sustainable development into every nation's policies and programmes and reverse the depletion of environmental resources; second, to reduce the loss of biodiversity and achieve a substantial reduction in the rate of such loss by 2010; third, to reduce the proportion of the universal population without sustainable access to clean and safe drinking water and basic sanitation by a half by 2015, and fourth, to achieve substantial improvement in the lives of a minimum of 100 million slum dwellers by 2020.

The MDGs emphasise that each nation's policies should be tailored to its particular needs, which is why most policy suggestions are rather general. The MDGs had a great international mobilising effect: they helped societies focus their attention on the fight against poverty and incentivised governments and societies in both developed and developing countries to raise the necessary resources to meet the agreed-upon goals. At the same time, it made evident that poverty eradication was a responsibility shared by all and thus a fundamental commitment of the international community. The MDGs favoured a sense of mission, therefore, which stimulated national efforts. Quantitative targets, associated with time-bound commitments, made possible both the follow up of progress as the concentration of national efforts.

However, the agenda also had important shortcomings. The goals simplified the international development agenda. Insistence on fighting extreme poverty and other basic needs helped leave out other important problems such as inequality, social fragmentation, youth unemployment, institutional fragility, crime, financial vulnerability or democracy consolidation at the same time as insufficient emphasis was placed on environmental sustainability (Deneulin and Shahani, 2009). Indeed, the MDGs do not capture all the elements necessary to achieve the ideals set out in the Millennium Declaration. Agriculture was not specifically mentioned in the MDGs even though most of the world's poor population are farmers. Not only do they exclude some of the most relevant topics, including the causes of poverty in some regions, but they make no reference to the policy framework, strategies and resources needed to achieve the formulated goals. The expectation was that results would automatically follow once resources were dedicated to the problems identified and, hence, little attention was placed on the regulatory frameworks that would help align private and public fund, both at the national and the international level, in order to achieve them.

This lack of strategy was also evident in the system developed by the WB to follow up and evaluate the achievements made, which favoured the selection of indicators and measurement tools set up by the most developed countries and thus biased against the least developed ones. This explains why the Agenda's results in Africa have been more modest compared with other regions.

Apart from this, the main shortcoming of the Millennium Agenda was the lack of financial resources to support it. Increased financing, linked to effective governance structures in low-income countries, can produce dramatic results. The increases in official development assistance (ODA) totalling only a few tenths of 1% of donor-country income, if properly directed and integrated into national PRSs, can substantially help reduce child mortality, hunger, lack of access to safe drinking water and sanitation, slum conditions in urban areas or lack of schooling.

The increased aid required to meet MDGs has been promised, but not yet delivered. In March 2002, governments worldwide adopted the Monterrey consensus (UN, 2003) at the International Conference on Financing for Development, which strengthened the global partnership needed to achieve MDGs. In essence, the international community recognised the need for a new partnership between rich and poor countries based on good governance and expanded trade, aid, and debt relief, especially to help finance the infrastructure and human capital needed to attract private investment. Donor countries committed themselves to providing the necessary resources by reaffirming their pledge to dedicate at least 0.7% of their income (GNP) to ODA, compared with the current average in the developed world of around 0.25%. With the combined donor-country GNP at roughly 30 trillion US dollars, 0.7% of GNP would be about $200 billion per year compared with the current aid flow of approximately $70 billion per year. The UN Millennium Project's findings show that the additional $130 billion per year would be more than enough to scale up the critical interventions needed to achieve the MDGs in well-governed, developing countries.

The Monterrey Consensus (2002) and the Doha Declaration (2008) also recognised the crucial role that ODA could play as an incentive for Foreign Direct Investment, as it could help mobilise private resources for some developing countries, thus alleviating pressure on public funding as well as stimulating international trade. The Doha Declaration focused its demands on two issues: first, on reforming international financial institutions and second, on improving the coordination of the financial, trade and development agencies that form part of the UN's multilateral system.

Apart from the key challenge of funding, the focus was placed on effectiveness in the management of resources and the implementation of goals. Effectiveness became a priority in 2002, together with funding, when the Monterrey Consensus was unable to raise sufficient resources for the implementation of the Agenda. The first High-Level Forum on Aid Effectiveness took place in Rome in 2003. Its conclusions focused on the need to harmonise the policies, procedures and operational practices of the institutions involved, in order to increase ODA's effectiveness in achieving the MDG (OECD, 2003). In Rome, donors agreed to coordinate their activities and reduce transaction costs for ODA-recipient countries. This harmonisation represented the first effectiveness-related commitment adopted by donor countries. Two years later, representatives from 35 donor countries, 26 multilateral agencies and 56 ODA-recipient countries met in Paris (2006) and decided to prioritise the recipient countries' development initiatives and strategies. In addition to appropriation by recipient countries, the Paris Declaration sets three additional effectiveness criteria. First, the alignment between aid donors and recipients; second, that result-oriented management should prevail over any other methodology and third, the idea of rendering all participants mutually accountable for

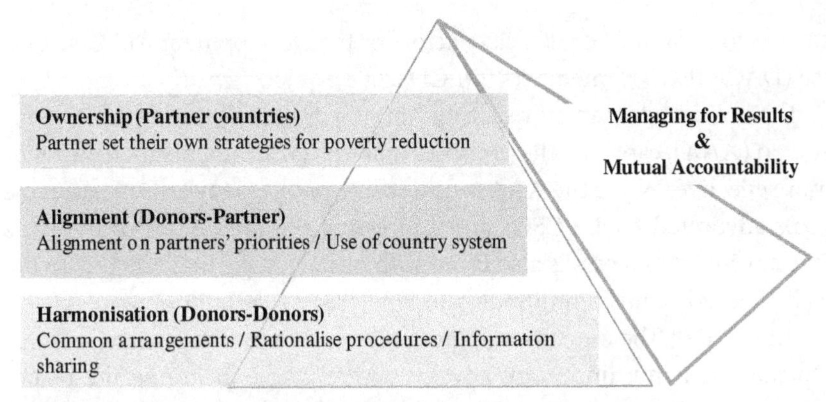

Figure 4.1 The Paris Declaration on aid effectiveness (2005).
Source: Harmonisation, Alignment, Results: Report on Progress, Challenges and Opportunities Joint Progress towards Enhanced Aid Effectiveness for the Paris H/L Forum, February 28-March 2, 2005 and other related documents.

their actions without asymmetries and thus regardless of their position and role as seen in Figure 4.1.

Three years of implementing the Paris Declaration and of a global reflection on the changing development landscape have provided some lessons and highlighted some limitations. European Civil Society Organisations (CSOs) welcome the intention of the signatories of the Paris Declaration to improve the technical and management dimensions of Aid Effectiveness, but also see the potential limitations of such a largely technical agenda, which could affect the means for the delivery of aid between governments (CONCORD, 2008). CSOs also highlight discrepancies between the principles of aid effectiveness and the practise and impact of aid on the ground. The commitments of the Paris Declaration are too focused on cost-effective aid delivery mechanisms, whereas CSOs propose to broaden and deepen the Paris Declaration process itself so that it reflects development effectiveness too. All these topics were exposed in the Third High-Level Forum on Aid Effectiveness in Accra, held in 2008 and attended by a broad coalition of over 380 CSOs from 80 countries. Progress was made in the Accra Forum as it fostered the inclusion of full-fledged associations with actors who did not participate in the design and definition of this Efficiency Aid Agenda, such as foundations, civil society actors or the Development Aid Committee (DAC/OECD)[m] members (OECD, 2008).

The political document resulting from the HLF3, the Accra Agenda for Action (AAA), captured the promises made by governments to make ODA more effective. While the AAA reflects some progress towards meeting the goals advocated for by CSOs and addressing the limitations of the Paris Declaration, its success is also faced with obstacles, particularly due to the lack of time-bound commitments and indicators to monitor progress (Better Aid, 2009). The outcomes of the AAA, therefore, reflect mixed results. On the one hand, important advances were made, including the recognition that gender equality, respect for human rights, and environmental sustainability are cornerstones for achieving enduring impact on the lives and potential of poor women, men and children and the inclusion of CSOs as full members of the Working Party on Aid Effectiveness (WP-EFF). In addition, the AAA sets out areas of action on the issues of predictability and transparency of aid flows, true ownership by CSOs and parliaments over aid decisions, reliance on the country systems of developing country governments rather than donor systems and a better and more efficient division of labour among donors. On the other hand, however, the AAA fails to put in place time-bound and monitorable commitments as well as indicators to measure progress in this regard. It also falls short in sufficiently addressing other key areas, such as decent work, policy conditionality, tied aid, mutual accountability and the reform of the aid governance system (*íbid*:4).

The fourth High-Level Busan Forum (2011) was developed in a different context than that of previous forums. only a minimum agreement was reached in relation to the Aid Effectiveness commitment, much was advanced in terms of consensus and participation. For the first time, China and Brazil, along with 161 other countries and 54 organisations, adopted "The Busan Partnership for Effective Development Co-operation" (OECD, 2011).

[m]Within the framework of Accra's progress, three new appointments took place in 2010, which complemented the effectiveness measures. First, the Bogot Declaration advanced the commitment regarding South–South cooperation in order to deepen the exchange of knowledge and mutual learning. Second, the Dali Declaration proposed to counteract the conflicts and consequent fragility of some countries through processes led by such countries themselves in the fields of peace keeping and state building. Finally, the Istanbul Declaration guided civil society organisations in their quest to be effective development actors.

The Busan Partnership document specifically highlights a set of common principles for all development actors that are key to making development cooperation effective (OECD, 2011).

- Ownership of development priorities by developing countries: Countries should define the development model that they want to implement.
- A focus on results: Having a sustainable impact should be the driving force behind investments and efforts in development policy making.
- Partnerships for development: Development depends on the participation of all actors, and recognises the diversity and complementarity of their functions.
- Transparency and shared responsibility: Development cooperation must be transparent and accountable to all citizens.

One of the main characteristics of the Busan Partnership is its emphasis on the role of aid as a complement to other sources of development financing, as aid on its own cannot break the poverty cycle (from aid effectiveness to effective development cooperation). As such, development cooperation should be a catalyst to mobilise resources to achieve development goals (OECD, 2011). The Busan partnership proposed the mobilisation of domestic resources in order to increase government resources and to do so, it urges development partners to fight more directly against corruption and tax evasion; to take a sturdier stance on strengthening national institutions under the leadership of developing countries; to build stronger relationships between development cooperation and the prívate sector by supporting the creation of a favourable environment for different partners and by fostering public-private partnerships and to share their experiences with actors involved in climate change financing, so they optimise the use of resources in a manner that is coherent with development policies.

The Busan Partnership also advanced in relation to the transparency of aid flows (IATI) to which the United States, Canada and the Inter-American Development Bank (IDB) and more than 50 members (donor and recipient countries, multilateral organisations and CSOs). Together with this initiative, a commitment was forged by the WB, the United Kingdom,

Switzerland, Spain, France, Estonia, Finland and the Netherlands to formalise an Open Aid Partnership that would work on mapping and publicly disseminating aid transfers so as to synchronise their agendas and improve their accountability. Finally, in Busan, a new treaty was reached regarding the Fragile States, in order to promote a new intervention agreement that would link aid to their particular needs for peace and institutional consolidation.

The Aid Effectiveness Agenda introduced the asymmetry between donors and recipients in the Aid debate, as a relevant topic in their actor's relations, but it was not enough especially from the South countries. Many of these countries were questioned for maintaining the same logics that defined the actions of the DAC/OECD donors and for not recognising the outputs of South-South cooperation nor the role of countries from the global South in the international development cooperation system. Currently, the Aid Effectiveness Agenda is not as relevant as it should be. Donor countries have not assumed their specific commitments, which have consequently not had an impact on improving the governance of cooperation. Moreover, the second meeting on the promotion of a Global Alliance for Effective Development, held in Mexico in 2014, did not prioritise the Aid Effectiveness Agenda for the post-2015 scenario (UN, 2014b). Linked to this missed opportunity, the European Commission (EC) emerges without leadership to promote this agenda and there is no one to relieve it.

The EC was very much involved in the design of the 2030 Agenda, stressing the importance of financial commitments and of the Aid Effectiveness Agenda. The European Consensus on Development, signed in 2005, was an agreement between Member States and the EC that reflected common values, objectives, principles and commitments in their respective development policies. In this regard, the consensus reflects a total agreement in reducing poverty (MDGs), but it goes one step further towards equitable globalisation, promoting development based on European values (respect for human rights, democracy, fundamental freedoms and the rule of law, good governance, gender equality, solidarity, social justice and effective multilateral action through the UN). Developing countries have the prime responsibility for their own development. But developed countries have a responsibility too. The EU, both at its Member State and Community levels, is committed to fulfilling its responsibilities. Working together, the EU is an important force for positive change. The EU provides over half of the world's aid and has committed to increasing this assistance, together with its quality and effectiveness. The EU is also the most important economic and trade partner for developing countries, offering specific trading benefits to developing countries, mainly to the least developed countries (LDCs) among them (EC, 2006).

The EU has adopted a timetable for Member States to reach the 0.7 % of GNP objective by 2015 (postponed to 2020), with an intermediate collective target of 0.56% by 2010,[n] while calling on partners to follow this lead. These commitments should see annual EU aid double to over EUR 66 billion in 2010. Further debt relief will be considered, as well as innovative sources of finance in order to increase the resources available in a sustainable and predictable way. At least half of this increase in aid will be allocated to Africa, while fully respecting individual Member States priorities' in development assistance. Resources will be allocated in an objective and transparent way, based on the needs and performance of the beneficiary countries, taking into account specific situations. Perhaps, the most important political commitment made by the EU, however, was the adoption of the Code of Conduct on Complementarity and Division of Labour in Development Policy (EC, 2006). The European Council recognised that reinforcing the complementarity of donor activities was of paramount importance for increasing aid effectiveness, and thus for a more effective and efficient development assistance, as one of the necessary conditions for the eradication of poverty in the context of sustainable development, including for timely achievement of the MDGs. Member States and the

[n]The Council conclusions adopted in May 2005 set out that Member States that have not yet reached a level of 0.51% ODA/GNI, undertake to reach, within their respective budget allocation processes, that level by 2010, while those that are already above that level undertake to sustain their efforts. Member States, which have joined the EU after 2002, and that have not reached a level of 0.17% ODA/GNI, will strive to increase their ODA to reach, within their respective budget allocation processes, that level by 2010, while those that are already above that level undertake to sustain their efforts. Member States undertake to achieve the 0.7% ODA/GNI target by 2015, whereas those that have achieved that target commit themselves to remain above that target; Member States that joined the EU after 2002 will strive to increase by 2015 their ODA/GNI to 0.33% (EC, 2006).

The Code of Conduct will guide policy and actions of the Member States and the Commission, and is embedded in the principles of ownership, alignment, harmonisation and management for results and mutual accountability of the Paris Declaration on Aid Effectiveness as well as the EU commitments set out in the European Consensus on Development (CE, 2007).

Commission will base their engagement in all developing countries on the principles set out in the Code of Conduct. The primary leadership and ownership in in-country division of labour should first and foremost lie in the partner country government. If such leadership and ownership need strengthening, the EU should promote it. In any case, the EU should always play an active role in promoting complementarity and division of labour (European Council, 2007). "All initiatives need to be open for other donors, build on existing processes whenever possible, and be readily transferred to the government whenever appropriate" (DAC/OECD, 2008). Environmental sustainability is included for the first time as a cross-cutting issue, to be integrated in all donor programmes areas, identified as a limitation key of growth, poverty reduction, equality, opportunity, security and with impact in every fields.

Nevertheless, the environmental considerations that re-included in the field of development focus on initiatives to ensure the sustainable management and preservation of natural resources, understood as a source of income and as a means to safeguard and develop jobs, rural livelihoods and environmental goods and services (EC, 2007). To this end, it will encourage and support national and regional strategies; it also took part in and contributed to European or global initiatives and organisations. Stronger support for the implementation of the UN Convention on Biological Diversity helped to halt biodiversity loss and promote biosafety and sustainable management of biodiversity. As far as desertification control and sustainable land management are concerned, the Commission focused on the implementation of the UN Convention to Combat Desertification. The European Consensus on Development improves through effective mainstreaming of sustainable land management issues in developing countries' strategies. With regard to sustainable forest management, the Community will support efforts on combating illegal logging and will pay particular attention to the implementation of Forest Law Enforcement, Governance and Trade. With regard to climate change, the Community focused its efforts on the implementation of the EU Action Plan on Climate Change in the context of development cooperation, in close collaboration with the Member States. Adaptation to the negative effects of climate change will be central in the Community's support to LDCs and small island development states. It also sought to promote the sustainable management of chemicals and waste, particularly by taking into account their links with health issues.

Next section analyses the relief assumed by the 2030 Sustainable Development Goals (SDGs) Agenda.

4.2 The International Sustainable Development Agenda (2015–2030) and the EU

In 2015, when the 2030 Agenda took over from the SDGs Agenda (UN, 2015a), the 2008 financial crisis entailed a different context. With a greater propositional and participatory ambition, the 2030 Agenda has sought to involve industrialised and underdeveloped countries in the sustainable development axis (under the climate change and development nexus) in an equal footing. Still, and as was the case with the MDGs Agenda, the 2030 Agenda lacks sufficient support from the international community. Without concrete commitments regarding its funding and with the Efficiency Aid Agenda lost in the urgencies of the financial crisis, the priorities of the global governance agenda are now focusing on protectionist issues and, in the case of the EU, on a global external action strategy in which security has taken centre stage.

The 2030 Agenda was adopted by the international community on September 2015, and includes in its nucleus the 17 SDGs and targets, which are still in force until 2030. Together with other international conferences and summits held in 2015, in Addis Ababa and in Paris, respectively, the international community has constructed a new framework within which all countries can collaborate to address common challenges. For the first time, the MDGs apply universally to all countries, and the EU has committed itself to fulfilling a pioneering role in their achievement. In 2016, the EC presented its ideas on a strategic plan to attain sustainable development in Europe and in the world, which included the proposal of a new Consensus. Since then, the European Parliament, Council and Commission coincide on a new collective vision on development policy that responds to the 2030 Agenda as well as other global challenges. Europe proposed a new European Consensus on Development in 2017 so that its development policy would lead collectively a new action plan to eradicate poverty and achieve sustainable development within the framework of the 2030 Agenda.

The new European Consensus on Development (European Commission, 2017) adopts for the first time a general and common framework for Europe's development aid that applies fully to all EU institutions and member states. The new consensus reiterates the primacy of poverty eradication within the aims of Europe's development policy and integrates the economic, social and environmental dimensions of sustainable development. It is a consensus-oriented towards people, the planet, prosperity, peace and good governance, which prioritises commerce and investment-oriented towards inclusive growth. As such, the consensus considers MDG 16 peace, security and governance, as an indispensable element for Democracy, and democracy itself as a priority in the achievement of sustainable development. With such consensus, the EU has adapted its development action to the 2030 Agenda, which is also a transversal dimension of the Global Strategy of the EU, which I will come back to.

In the common framework of the European Consensus on Development, commitments focus on three main areas: the interconnection between development and external policy, a global vision on the means of execution and on a better integration between actors. First, the agreement recognises the numerous interconnections between development, peace and security, humanitarian aid, migration, the environment and climate, as well as other transversal issues such as youth, gender equality, mobility and migration, sustainable energy and climate change, investments and commerce, governance, democracy, the rule of law and human rights. Second, the new consensus applies, in addition, a global vision to the means of execution, which combines traditional development aid with other resources as well as with good policies and strengthened approach towards action coherence, remembering that the development aid of the EU has to always be considered within the context of the own efforts of Europe's partner countries. The consensus lays the foundations for the EU and its member states to participate in more innovative forms of funding development, strengthening the private sector and mobilising more national resources for development. Third, the EU and its member states commit themselves to the creation of better-adapted associations with a broader array of stakeholders, including civil society, and with partner countries in all phases of development. In this remit, what is sought is better collaboration, considering all the respective comparative advantages.

However, is the European Consensus on Development a response from the EU to global tendencies and challenges or is it an adaptation of its external action to the 2030 Agenda on Sustainable Development? To answer this question, we must revise the commitments that the EU has acquired since the 2030 Agenda in relation to two interrelated international agendas: the Third Conference on Funding for Development held in Addis Ababa in July 2015 and the Political Conference on Climate held in Paris in December 2015.

The Third Conference on Funding for Development (Addis Ababa (UN, 2015b)) took place without the application of the acquired commitments nor the agreed upon reforms. Important commitments were created, therefore, but without any real support from the international community. The most relevant topics addressed focused on the persecution of illegal funding, the ratification of the UN Convention against Corruption, especially in extracting industries, the widening of international cooperation on taxing issues, and so on, the renewal of the Monterrey Consensus as a complement to national development initiatives, the international flows of private capital and the recognition of international commerce as a decisive factor in development (the need for a greater effort in multilateral commercial negotiations within the remit of the WTO) (UN, 2015b). In Addis Ababa, the commitment to allocate 0.7% of the GNP to development aid by the most industrialised countries is ratified, as well as that of 0.15% and 0.20% of the least advanced countries (UN, 2015c). A commitment that the EU has joined, but collectively and that in 2017, was postponed to 2020.

With the results of the report of the Intergovernmental Expert Group on Climate Change, which resulted from the Paris Conference on Climate, 195 countries signed the first globally binding agreement on climate. The latter ratified the commitment to adopt strong mitigation measures by means of the first Global Agreement on greenhouse gas emissions in light of their direct effect on global warming. To avoid climate change, the UN Framework Convention on Climate Change (UCFCCC) establishes a global action plan that sets the limit of global warming below 2°C, limiting the increase to 1.5°C, which would considerably reduce the risks and impact of climate change. Among the measures proposed, the most important include a global reduction in the costs of consumption of between 1 and 4% by 2030 and between 2 and 6% by 2050, as well as the creation of a Green Fund for Climate that will allow achieving the commitments with developing countries through international aid (UNFCCC, 2015).[p] The national action plans against climate change presented by the countries in the Paris Conference, however, were not just insufficient to maintain global warming under 2°C, but it was not until the following year in Marrakech (2016) that the difference between industrialised and developing countries was made effective; a difference that had been on the agenda since Political Conference on Climate held in Varsovia in December 2013(UNFCCC, 2016). In this sense, measures such as appropriate funding flows, a new technological framework and a better framework for the creation of capabilities to support the action of developing and vulnerable countries in line with their own national objectives, were considered.

The EU formally ratified the Paris Agreement in 2016, which allowed it to come into force on 4 November of that same year. It also joined the objective of raising 100,000 US dollars a year in 2020 and increasing it until 2025 (EC, 2016). The EU has been at the forefront of the international efforts to reach a global agreement on climate. In light of the limited participation in the Kyoto Protocol (UNFCCC, 1998), and the absence of any agreement in the Copenhagen Summit (2009) (UN, 1998, 2009), the EU fostered the creation of a wide and ambitious coalition of developed and developing countries which prefigured the good results of the Paris Conference. In March 2015, the EU was the first great economy to present its expected contribution to the new Agreement and in 2016 was already taking measures to reach its objective of reducing emissions by a minimum of 40% by 2030.

The EU not only took up a leading position. In the financial realm, the objective of allocating 0.7% of the GNP, which the EU committed to in 2015 within the context of the MDGs agenda, was postponed to 2020. Political commitments need to be accompanied with the resources that will make them viable and the EC counts only with 15% of the total EU ODA, for which the material commitment of member states becomes necessary.

[p]In the COP 21 of the strategic commitments regarding development cooperation focus on assigning a minimum of 50% to the adaptation of the especially vulnerable countries, among them the least advanced ones, small insular countries and the countries of Africa (UNFCCC, 2013).

Brexit will entail the loss of an important partner in quantitative terms (also in terms of the great capacities of the British development agency, DFID). Together with Germany, which this year has reached the 0.7% objective for the first time, other large and medium countries will have to increase the resources necessary to achieve this objective. It is a pending task in order to gain credibility (Ayuso, 2017). Leadership will also have to be demonstrated by strengthening the commitment of the Paris Agreement regarding climate. In light of the United States' withdrawal (2017) (Milman et al., 2017), the EU should have assumed a greater leadership in the use of clean energy and the reduction of contaminating emissions, and should have searched for alliances that would also contribute to the achievement of the MDGs, the commitments of Rio+20 both inside and outside its borders, as well as helping countries with less resources to join the global efforts against the temptation to follow the United States' short-term vision.

In addition to institutions, resources and technology, however, the leadership of the EU should be reflected in the consistency between discourse and principles with practise. Policy Coherence for Sustainable Development (PCSD) should be a substantial element of Europe's influence in the international development agenda, as the European Consensus of Development of 2017 itself states. The la European confederation of Relief and Development Non-Governmental Organizations (CONCORD),[q] however, has highlighted the excessive instrumentalisation of aid by security objectives, as is the case with the conditioning of aid to the return of asylum-seekers. The incoherence between the discourse and practise of the EU in the context of the "refugee crisis" has undermined the moral leadership of the EU in the 2030 Agenda. If the new consensus cannot resolve such incoherence, it might work to improve the efficiency of some policies, but not to lead the SDGs Agenda.

[q]CONCORD member organisations: 28 national associations, 24 international networks and 3 associate members that represent more than 2,600 NGOs, supported by millions of citizens across Europe. It is the main interlocutor with the EU institutions on development policy. We aim to strengthen the impact of European development NGOs vis-à-vis the European Institutions and to positively influence the European Development policies for a fair, just and sustainable world (CONCORD, 2018).

4.3 Challenges between the EU Global Security Strategy and SDGs Agenda

The SDGs Agenda implies a revision of all the areas of the PCSD, including the EU's 10 priorities, as well as committing to global governance. Yet, the interpretation and limits of such governance are not considered in the same way in all areas nor by all of Europe's partners, as can be appreciated in current events (nationalisms and extremisms). The European Consensus of Development, which tries to guide the approach of member states and the EU regarding development cooperation and the SDGs until 2030, has very much in mind the external and security policies of the EU, and, in combination with the Global EU Strategy adopted in 2016, presents a potential blueprint for a new European vision where the objectives and interests of its external action can become the main challenge.

The weight of the EU's regional dimension in such political guidelines turns the PCSD into a priority. Such a concept, which is increasingly recurrent in the debate about the efficiency of aid not only allows for the integration of both approaches: sustainable development and the reduction of poverty. It also implies the transversal application to all the measures adopted by European institutions of the SDGs, as an integral commitment for all the decisions for which the EU is responsible as well as for its coordinated action with states. The achievement of the PCSD at a European level is not exempt from difficulties. Both in theory and practice, the parameters and visions are multiple in an approach that has different priorities.

There are those, on the one hand, who consider it imperative to integrate the concept of PCSD in order to manage globalisation by means of the creation of collective action that allows for the common struggle against the new transnational challenges. On the other hand, there are those who consider a transversal agreement on sustainable development affecting all areas of public action sufficient, or those who understand the PSCD as the best way to detect contradictions and inconsistencies between governmental initiatives in different domains (Ashoff, 2002). We also find a less ambitious approach where the PSCD rests solely on the promotion of synergies and complementarities, which serve little to correct incompatibilities or modify inconsistencies (Millán, 2014).

The Treaty of Lisbon (2007) imposes the obligation of incorporating the objectives of development policies into other policies of the EU that could affect developing countries (Art. 130). An obligation that was already included in the European Consensus of Development (2005) and that urged member states, in the first place, to improve their practises and policies in the field of PSCD; it urged the Council, second, through the integration of the different aspects related to development into the work of sectoral groups, and third, the Commission, by means of fostering an analysis of the impact of policies from a development perspective (EC, 2006). All of the EU's policies, ultimately, are called to support, or at least not hinder the objectives of development, democracy, good governance, security and human rights beyond European frontiers (EC, 2011).

One thing is theory, however, and another, practice. Priorities seem to be different if we consider the destiny of European policies. When these emerge regarding inside its borders, the EU seems to have worsened the rights of its own citizens. The commitment to growth and macroeconomic stability has given way to an involution in those rights, which, however, can be seen in the policies that regard outside of its borders. This is reflected in its Agenda for Change (2011), for example, particularly in its proposal for a common response of the Union and its member states with thematic programs based on synergies between general interests and the eradication of poverty in partner countries. A strategy where we can appreciate the embryonic attempts to attend differently to the different situations in such countries and make more efficient decisions regarding policies, levels of help, assistance regimes and the use of financial resources (EC, 2011).

In turn, such collective programming of the aid of the EU and its member states, which should reduce the problem of fragmentation and increase its impact beyond the strategies of each partner country, has not been completely efficient. Given that development aid is a shared competence (Art. 4 TFUE), the Union's action has to coexist with that of member states. Even if the subsidiarity principle (Art. 5.3 TUE) allows the EU to avoid overlaps and guarantee the complementarity of actions, the truth is that the work dedicated to the promotion of development undertaken by European institutions and member states is characterised by a significative gap between real actions and political commitments.[r]

In this sense, and despite the fact that the institutions, instruments and qualified personnel of the Union has entailed an advancement, the role of the EU and its adaptation of the SDGs Agenda has to be better linked to the PSCD and its new frameworks of development cooperation. Only in relation to external policy, the action of the EU and its member states denote an important democratic deficit and a lack of promotion of human rights. Despite having a charter of Fundamental Rights that includes their defence and extends their application to other policies of the EU (its external and security policy, financial cooperation, trade and immigration), and with a compulsory clause for all its agreements, where the basis of European relations are set within the respect of human and democratic rights, its participation in the consolidation of some dictatorships can be attested. Between 1998 and 2005, the EU signed association agreements with Libya, Egypt, Algeria, Tunisia, Jordan, Israel and Morocco, the respect of democratic principles and fundamental rights being a fundamental element in them. Similarly, we can confirm the superposition of security objectives in other bilateral relations of member states. The Friendship, Society and Cooperation Treaty signed between Silvio Berlusconi and Muammar Gadafi engaged the latter in the fight against terrorism, illegal immigration and drug trafficking in exchange form 3,600 million euros.[s]

In this line, the arms trade also entails a great challenge. In the international domain, the Arms Treaty promoted by the UN[t] establishes since 2014 a set of common standards for the 91 signatory countries among which member states are included. In the national domain, each member state has its own legislation to control its export and since 2008, in addition, with a common European position on the Export of Military Technology and Equipment,[u] which links the responsibility of the exporting country

[r] Of 164 new policies of the EU, only 7 have analysed their impacto on developing countries.

[s] As compensation for the Italian occupation in the 20th century.

[t] The Arms Trade Treaty (UNODA, 2013), regulated conventional arms trade and entered into force on the 24 December 2014.

[u] COUNCIL COMMON POSITION (2008) *Common Position defining common rules governing control of exports of military technology and equipment*, 2008/944/CFSP of 8 December 2008.

to the final use of the weapons; a responsibility that has to respect International Humanitarian Law. Regardless, between 2000 and 2009 the export of the EU in matters of defence increased an 871% towards other OECD countries as well as developing countries with latent conflicts (Colombia, Morocco, Saudi Arabia, Rwanda, India, Ghana, Turkey, Israel, Pakistan and Sri Lanka). As has been described, to advance in the PSCD is a necessary condition for the EU and its member states. In order to comply with the SDGs Agenda, it is not just necessary to understand sustainable development, but also to pace it with the different rhythms and interests of all countries in a world that has not yet specified its commitments to reduce the impact of its life style on climate.

Conclusions

The Global Strategy on Foreign and Security Policy of the EU (2016) establishes as a priority, by means of development, diplomacy and the common defence and security policy, to strengthen the capacities of partner countries in creating security within the framework of the rule of law, as well as the resilience of the Union's neighbouring regions. An internal scenario that imposes itself on external action.

On the other hand, the EU has been and is present, though in different degrees, in the International Development Cooperation policy. Formally, it has not just participated in the commitments of the agendas already mentioned, but has also established more precise ones, which apparently linked a greater responsibility on behalf of the EU with development and governance. Europe's support to the International Development Cooperation policy has overcome, not without distortions, the weight of the financial crisis and the isolationist behaviour of other relevant actors. At the same time, the EU is wasting the opportunity created by the distancing of China and the United States from the implementation of the SDGs Agenda. All that political performance should translate into a European leadership that could advance, without fissures, and less rhetoric, a form of governance based on sustainable development, responsible and coherent with the commitments acquired and, most importantly, the initiative of constructing common global goods to address climate change.

References

Ashoff, G. (2002) *Improving Coherence between Development Policy and Other Policies*. The Case of Germany. German Development Institute. Briefing paper.

Ayuso, A. (2017) *¿Puede el Nuevo Consenso de Desarrollo Europeo hacer de la UE un líder de la Agenda n.d.? Nº Opinion 486, CIDOB*, available at: https://www.cidob.org/en/publications/publication_series/opinion/seguridad_y_politica_mundial/puede_el_nuevo_consenso_de_desarrollo_europeo_hacer_de_la_ue_un_lider_de_la_agenda_2030

Better Aid (2009) *An Assessment of the Accra Agenda for Action from a Civil Society Perspective*, available at: https://www.fingo.fi/sites/default/tiedostot/better-aid-assessment-aaa-oct-2009.pdf

CONCORD (2008) *Contribution to the "EU Aid Effectiveness Roadmap to Accra & Beyond" - CSO and the principles of the PD. Enlarged Taskgroup on CSO Effectiveness in Development Cooperation, January*, available at: https://diplomatie.belgium.be/sites/default/files/downloads/EU_CSO_contribution_Roadmap_Accra.pdf

CONCORD (2018) *Who We are*, available at: https://concordeurope.org/

Council of the European Union (2007) *EU Code of Conduct on Complementarity and Division of Labour in Development Policy*. 9558/2007. DEVGEN 89 ACP 94 RELEX 347, available at: http://register.consilium.europa.eu/doc/srv?l=EN&f=ST%209558%202007%20INIT

Council of the European Union (2008) *Posición común 2008/944/PESC de 8 de diciembre*, available at: http://eur-lex.europa.eu/legal-content/ES/TXT/?uri=CELEX%3A32008E0944

Council of the European Union (2011) *Operational Framework on the Effectiveness of Development Aid*, available at: http://data.consilium.europa.eu/doc/document/ST-18239-2010-INIT/en/pdf

Deneulin, S. and Shahani, L. (2009) *An Introduction to the Human Development and Capability Approach Freedom and Agency*.Earthscan in the UK and USA: London, available at: https://idl-bnc-idrc.dspacedirect.org/bitstream/handle/10625/40248/IDL-40248.pdf

EEAS (2017) *From Shared Vision to Common Action: Implementing the EU Global Strategy Year 1 A Global Strategy for the European Union's Foreign and Security*. European External Action Service, Brussels. available at: https://europa.eu/globalstrategy/sites/globalstrategy/files/full_brochure_year_1.pdf

EU (2006) *European Development Consensus. 2006/c 46/01 (DO C 46, 24 de febrero)*, available at: https://goo.gl/MYomU3

EU (2007) *Treaty of Lisbon amending the Treaty on European Union and the Treaty establishing the European Community, signed at Lisbon, 13 December. (2007/C 306/01)* available at: https://eur-lex.europa.eu/legal-content/EN/TXT/?uri=CELEX%3A12007L%2FTXT

EU (2012) *Treaty on European Union and the Treaty on the Functioning of the European Union 2012/C 326/01*, available at: https://eur-lex.europa.eu/legal-content/EN/ALL/?uri=CELEX:12012E/TXT

EU (2016) *EU Global Strategy*, available at: https://europa.eu/globalstrategy/en

European Commission (2006) *The European Consensus on Development.* Directorate-General for Development. Luxembourg, Office for Official Publications of the European Communities, available at: https://ec.europa.eu/international-partnerships/system/files/publication-the-european-consensus-on-development-200606_en.pdf

European Commission (2011) *Communication from the Commission to the European Parliament, the Council, the European Economic and Social Committee and the Committee of the Regions. Increasing the Impact of EU Development Policy: an Agenda for Change.* {SEC(2011) 1172 final}, available at: https://eur-lex.europa.eu/legal-content/EN/TXT/PDF/?uri=CELEX:52011DC0637&from=en

European Commission (2016) *The European Parliament has Approved the Ratification of the Paris Agreement by the European Union Today*, available at: https://ec.europa.eu/clima/news/articles/news_2016100401_en

European Commission (2017) *The New European Consensus on Development 'Our World, Our Dignity, Our Future' Joint Statement by the Council and the Representatives of the Governments of the Member States Meeting Within the Council, the European Parliament and the European Commission*, available at: https://op.europa.eu/en/publication-detail/-/publication/5a95e892-ec76-11e8-b690-01aa75ed71a1

IMF (1999) *The Poverty Reduction and Growth Facility (PRGF)—Operational Issues*, available at: https://www.imf.org/external/np/pdr/prsp/poverty2.htm

Millán, N. (2014) Reflexiones para el estudio de la coherencia de políticas para el desarrollo y sus principales dimensiones. *Papeles 2015 y más,* 17, 1–17.

Milman, O., Smith, D. and Carrington, D. (2017) Donald Trump Confirms US will Quit Paris Climate Agreement World's Second Largest Greenhouse Gas Emitter will Remove itself from Global Treaty as Trump Claims Accord 'Will Harm' American Jobs. *The Guardian*, available at: https://www.theguardian.com/environment/2017/jun/01/donald-trump-confirms-us-will-quit-paris-climate-deal

OECD (2003) *HLF1: The First High Level Forum on Aid Effectiveness, Rome Declaration on Harmonisation*, available at: http://www.oecd.org/dac/effectiveness/31451637.pdf

OECD (2005) *The Paris Declaration on Aid Effectiveness: Five Principles for Smart Aid*, available at: https://www.oecd.org/dac/effectiveness/45827300.pdf

OECD (2008) *The Accra Agenda for Action (AAA)*, available at: https://www.oecd.org/dac/effectiveness/45827311.pdf

OECD (2011) *Fourth High Level Forum on Aid Effectiveness, Bussan*, available at: http://www.oecd.org/dac/effectiveness/49650173.pdf

Pisano, U., Endl, A. and Berger, G. (2012) *The Rio+20 Conference 2012: Objectives, processes and Outcomes*. European Sustainable Development Network. Vienna University of Economics and Business Franz Klein Gasse 1, A-1190 Vienna, Austria, available at: https://www.sd-network.eu/quarterly%20reports/report%20files/pdf/2012-June-The_Rio+20_Conference_2012.pdf

Ryland, T. and Samuel, H. (2018) *What Was the U.S. GDP Then?* Measuring Worth.

Sachs, J. (2015) *The Age of Sustainable Development*. Columbia University Press: New York.

UN (1987) *Report of the World Commission on Environment and Development: Our Common Future*, available at: http://www.un-documents.net/our-common-future.pdf

UN (1992) *United Nations Conference on Environment & Development Rio de Janerio, Brazil, 3 to 14 June 1992*, available at: https://sustainabledevelopment.un.org/content/documents/Agenda21.pdf

UN (1998) *Kyoto Protocol to the United Nations Framework Convention on Climate Change*, available at: https://unfccc.int/resource/docs/convkp/kpeng.pdf

UN (2000) *Resolution Adopted by the General Assembly (A/55/L.2)]*. United Nations Millennium Declaration: New York, available at: https://undocs.org/A/RES/55/2

UN (2002) *Report of the World Summit on Sustainable Development Johannesburg, South Africa, 26 August 4 September*. United Nation Publications: New York, available at: http://www.un-documents.net/aconf199-20.pdf

UN (2003) *Monterrey Consensus of the International Conference on Financing for Development. The Final Text of Agreements and Commitments Adopted at the International Conference on Financing for Development Monterrey, Mexico, 18–22 March 2002*. United Nations Department of Economic and Social Affairs Financing for Development Office: New York, available at: https://www.un.org/esa/ffd/wp-content/uploads/2014/09/MonterreyConsensus.pdf

UN (2008) *Doha Declaration on Financing for Development: outcome document of the Follow-up International Conference on Financing for Development to Review the Implementation of the Monterrey Consensus. Doha, Qatar, 29 November-2 December*. A/CONF.212/L.1/ available at: http://archive.ipu.org/splz-e/finance09/doha.pdf

UN (2009) *The Conference of the Parties, Takes note of the Copenhagen Accord of 18 December 2009*, available at: http://unfccc.int/files/meetings/cop_15/application/pdf/cop15_cph_auv.pdf

UN (2012) *United Nations Conference on Sustainable Development, Rio+20. Resolution Adopted by the General Assembly on 27 July 2012 [Without Reference to a Main Committee (A/66/L.56)] 66/288. The Future We Want.*

United Nations Publications: New York, available at: http://www.un.org/ga/search/view_doc.asp?symbol=A/RES/66/288&Lang=E

UN (2014a) *The Arms Trade Treaty (ATT)*, available at: https://www.un.org/disarmament/convarms/att/

UN (2014b) *The First High-Level Meeting (HLM1) of the Global Partnership for Effective Development Co-operation Marked a Major Milestone in the Global Fight Against Poverty. Secretaría de Relaciones Exteriores / Agencia Mexicana de Cooperación Internacional para el Desarrollo*, available at: http://effectivecooperation.org/wp-content/uploads/2015/01/MEMORIA-FINAL.pdf

UN (2015a) *Sustainable Development Goals.* United Nation Publications: New York, available at: https://www.un.org/sustainabledevelopment/development-agenda/

UN (2015b) *Report of the Third International Conference on Financing for Development Addis Ababa 13–16 July 2015.* United Nation Publications: New York, available at: http://www.undocs.org/A/CONF.227/20

UN (2015c) *69 Session. Resolution adopted by the General Assembly on 27 July 2015 [without reference to a Main Committee (A/69/L.82)] 69/313. Addis Abeba Action Agenda of the Third International Conference on Financing for Development*, available at: http://www.undocs.org/A/RES/69/313

UNFCCC (2013) *UNFCCC COP 19 and CMP 9 and the Subsidiary Bodies will Convene in Warsaw, Poland*, available at: http://sdg.iisd.org/events/conference-of-the-parties-to-the-unfccc/

UNFCCC (2015) *Paris Agreement*, available at: https://unfccc.int/sites/default/files/english_paris_agreement.pdf

UNFCCC (2016) *Marrakech Climate Change Conference – November. Marrakech Action Proclamation for Our Climate and Sustainable Development*, available at: https://unfccc.int/files/meetings/marrakech_nov_2016/application/pdf/marrakech_action_proclamation.pdf

United Nations(2008) *The Arms Trade Treaty (ATT), Regulated Conventional Arms Trade and entered into force on the 24 December 2014*, available at: https://www.un.org/disarmament/convarms/att/

UNODA (2013) *The Arms Trade Treaty*, New York. 2 April 2013. Available at https://treaties.un.org/pages/CTCs.aspx

World Bank (2018) *Heavily Indebted Poor Country (HIPC) Initiative*, available at: http://www.worldbank.org/

CHAPTER 5

A European Action Plan for Circular Economy

Elena Bulmer Santana

The circular economy should be a central political project for Europe, as it offers the potential to set a strong perspective on renewed competitiveness, positive economic development, and job creation.

–Ida Auken, Member of Parliament, Denmark[a]

5.1 A European Circular Economy

To develop a sustainable, low-carbon, resource-efficient, competitive economy, Europe is striving for a more Circular Economy (CE), "where the value of products, materials and resources is maintained in the economy for as long as possible and the generation of waste is minimised" (European Commission, 2017a). Furthermore, CE also proposes an economic model that will reduce greenhouse gases to the levels that were accorded at the Paris Climate agreements to keep "a global temperature rise this century well below 2°C above pre-industrial levels and to pursue efforts to limit the temperature increase even further to 1.5°C" (The Paris Agreement, 2019).

[a]Taken from Ellen MacArthur Foundation report titled, *Growth within: A circular economy vision for a competitive Europe* (2015).

CE has been gaining growing attention among both practitioners and scholars (Kirchherr et al., 2018).

CE needs to recognise and deal with three main issues, which are resource scarcity, environmental impact and in parallel the enhancing of economic benefits (Holton et al., 2010; Ritzén and Sandström, 2017), while providing solutions that convert environmental challenges into economic opportunities (Costea-Dunarintu, 2016). It is a model very different from the linear traditional economy model, which we have been so much dependent upon to date, that is based on a "take-make-consume-throw away" approach to resources. There are manifold reasons for moving to this new type of economy according to the European Commission (EC) (European Commission, 2015): Protect business from the scarcity of resources (i.e., and thereby helping to protect against supply and price challenges); create new and creative business opportunities; develop more efficient ways of producing and consuming; create new jobs at all levels; reduce the probability of irreversible environmental damage (i.e., with regard to biodiversity, climate, air pollution, etc.); provide savings in energy usage (European Commission, 2016b) and provide a greener public and social image for businesses through green marketing for their services and products (Korhonen et al., 2018).

To date, many businesses, industries and other economic stakeholders have demonstrated their support for CE, however, CE is still at its initial phases in the EU. The actual implementation of CE is not easy and is facing a number of different barriers among the different EU Member States (i.e., which will be described later on) (Kirchherr et al., 2018).

In December 2015, the EC adopted the EU Action Plan for CE (European Commission, 2015). The main objective of this Action Plan is to establish a long-term CE strategy for the EU. Through it, the EU and EU Member States would be putting into action global commitments entered into, especially the UN 2030 Agenda for Sustainable Development (European Commission, 2015) and the COP21 Paris Agreements. One of the main objectives of the EU CE Package is to improve European competitiveness by helping to protect industries from the potential risk of resource scarcities and consequent pressure on prices. The measures in the package will also help to foster new business opportunities, as well as innovative approaches to production and consumption (Milios, 2018).

What is worth highlighting at this point in the discussion is the importance of the long-term involvement of stakeholders at all levels; from Member States, regions and cities, to business and citizens. The uncertainty that is present in this whole transition process towards a more CE could potentially be alleviated through a greater degree of communication regarding this issue at all levels. The generation of greater awareness regarding the safety of recycled materials will create trust and in turn encourage demand for them, and thereby promote change rather than foster resistance to it.

5.2 The Four Stages of CE

CE is important throughout a product's life cycle; from the beginning of a product's life (i.e., from the point that the raw or primary materials are extracted from their source) to the point at which they become waste. These stages are described below in more detail.

5.2.1 *Stage 1: Production*

The start of a product's life cycle starts with the product design. Through better design, it is possible to make products more durable, or easier to repair, upgrade or remanufacture. Improved designs will render products more durable and easier to reuse, repair, upgrade and remanufacture, thereby making them operational for longer and benefitting users, as repair and replacement cost will be lower (Milios, 2018). It is important to ensure that their design allows for products to be capable of being disassembled and adapted for reuse (Sacchi et al., 2018). There are two main elements here of high importance that need to be considered that are access to spare parts at a reasonable cost, as well as access to necessary information (Milios, 2018). At present, however, some manufacturers fail to disclose specific information regarding their products, thus creating difficulty for commonality in the design of specific parts for similar products, and thereby limiting and increasing the cost of their reparability (Maitre-Ekern and Dalhammar, 2016).

With better design, it will be easier to disassemble and dismantle products so to recover valuable components, thereby reducing the use of raw materials. In this regard, the EC has developed the Ecodesign Directive, the objective of which is to improve the efficiency and environmental performance of energy-related products (EU Directive, 2009), by facilitating the dismantling, reusing and recycling of electronic displays such as computer or television screens.

The Ecodesign Directive (i.e., in its Ecodesign 2015–2017 working plan) developed energy-saving Ecodesign for household appliances, information technologies or engineering. These requirements are in turn implemented via product-specific regulations, which can be directly applied in all EU countries. (European Commission, 2018c). The Ecodesign is complemented with the Energy Labelling Regulation through the development and implementation of mandatory labelling requirements. Both aforementioned regulations are further complemented with the European harmonised standards, which also need to be complied with (i.e., with regard to its technical specifications), thus allowing the use of the "CE" mark and for the product to be sold in Europe (European Commission, 2018c). The prime objective of these regulations is to strengthen Europe's competitiveness, job creation and economic growth in a sustainable manner (European Commission, 2016a). The Ecodesign and energy labelling framework is considered to be one of the most effective EU instruments and is expected to curb energy savings by 50% by 2020, equating to 175 Mtoe per year, while generating 55 billion euros extra of revenues for industry, creating 800,000 new jobs and reducing CO_2 emissions by 320 million tonnes annually (European Commission, 2016a).

The recent Ecodesign Working Plan is meant to contribute to the Commission's CE Action Plan, with regard to (European Commission, 2016a) resource efficiency, reparability, recyclability and durability.

Once the products have been designed smartly, resources must then also be used just as smartly and efficiently in order to yield maximum benefit for the business and reduce waste generation. The efficient use of primary raw resources in the different production processes must be targeted here in order to avoid their becoming scarce and thereby reducing price pressure on their supply which might affect the profitability or the survival of the business in question. Furthermore, special attention and consideration must also be given to both the social and environmental implications both in the EU and outside the EU. Raw material sustainable sourcing comes into play here. One of the indicators of the CE Monitoring Framework (described below in more detail) is "Green Public Procurement" aimed at ensuring that the procurement performed by both public and private institutions is done sustainably. As major consumers, the public authorities of Europe may utilise their purchasing power to select environmentally sustainable products, therefore making a considerable contribution towards sustainable consumption and production. Green public procurement is very much needed to help Europe become a resource-efficient economy. It may also encourage appreciable demand for specific sustainable products for which there would not normally be a demand. Such would be the case for secondary (i.e., recycled) materials and for eco-innovative products, thereby stimulating at the same time eco-innovation. However, for the time being, green procurement still needs to be further promoted among European authorities as to date it is not very widespread. Although regulations and standards have been developed at the national level in some European countries, there is still the challenge of standardising green procurement requirements between Member States in order to stimulate the development of a European single market for sustainable goods and services (European Commission, 2018d). This lack of standardisation (i.e., with the lack of explicit regulations) only creates barriers, especially when we take into consideration the multiple numbers of stakeholders involved in green procurement processes, such as national agencies, citizen organisations, market actors and procurement agencies just to mention a few (Günther and Scheibe, 2006; Milios, 2018). Furthermore, there is a need for research as regards to eco-design standards for repair and durability based on consumer preferences (Milios, 2018).

The development and implementation of CE will contribute towards the reduction of greenhouse gases, thereby complying the 2015 Paris Climate targets by enabling the meeting of material needs without surpassing the currently available carbon budget (Bourguignon, 2016). The EC estimates that 4%–6% of greenhouse gas emissions could be reduced thanks to EC. It is also estimated that EU could cut the industrial emissions by 56% and create a number of different opportunities: (1) Materials recirculation opportunities, (2) Product materials efficiency and (3) New circular business models (Enkvist & Klevnäs, 2018).

Finally, getting the EU to lead technological advances with regard to CE is one objective of the EU's CE Action Plan. The EU presently has a number of different funding schemes (i.e., described below in further detail) that promote and support research and development in CE, such as the Horizon 2020 and the EU Cohesion Fund (European Commission, 2018f).

CE, as previously mentioned, includes the notions of reduction, reuse and recycling, which presents a context in which there is a need for optimised networks (Yu et al., 2013; Sacchi et al., 2018). Furthermore, it is also important to fund innovative industrial processes such as industrial symbiosis, an example of an optimised network, where the by-products of one process become the input to another, otherwise known as a "zero waste economy." A wonderful example of industrial symbiosis is Kalundborg in Denmark, where one finds a compendium of different co-located companies optimising resources through the symbiotic exchange of materials such as water and steam (Brings Jacobsen, 2006). According to the Ellen MacArthur Foundation, CE is defined as a closed-loop industrial system that is restorative and regenerative by intention and design (MacArthur, 2015).

5.2.2 *Stage 2: Consumption*

It is essential for there to be a demand for sustainable products. Consumers are the main drivers in the transition towards a CE in Europe (Milios, 2018). The choices of millions of these consumers can shape the direction of CE in Europe. Often these choices are very much affected and moulded by the information that consumers are presented with, as well as the range and prices of existing products (European Commission, 2015). However, often consumers may be bombarded by many different green labels, some of which lack accuracy and clarity, and this may over time cause consumers to lose trust in sustainability labelling. The Commission is, therefore, working on turning the situation around and making green labelling more trustworthy by ensuring that there is a better implementation of the rules in place (European Commission, 2015). An example of this is the improved labelling for the energy performance of household electrical appliances, through the above-described Energy Labelling Regulation. With adequate information, consumers will consequently be able to select the most energy-efficient and the most environmentally friendly appliances. Another important point to highlight is the problem of programmed obsolescence. The Action Plan aims to detect and deal with such practices.

Moreover, a tool that is very important for the CE context is the EU Ecolabel, which has over several decades supported sustainable production and consumption. To date in Europe, over 40,000 products carry the EU Ecolabel. Products that are "eco-labelled," contribute to sustainable development throughout their life cycle thereby ensuring resource efficiency, durability and repairability. The EU Ecolabel tool has for the last 26 years become a key legislative element in the promotion of greener products and materials in Europe. In 2017, the European countries with the most EU Ecolabel licenses were France, Italy and Germany (European Commission, 2017b).

At present, there is only a 6% demand for recycled products (also known as secondary materials). The development of the markets for these materials, however, is fundamental to improving this current figure (European Commission, 2015). Increasing this percentage is essential as the reuse and repair of products is critical for the EU, as it will first reduce wastage, and second support its job agenda, creating millions of new employment opportunities and it is, therefore, essential to extend consumer trust in recycled products through public awareness campaigns. Furthermore, there is an innovative model of consumption, which is the collaborative economy, where different products or infrastructure are shared among different consumers (European Commission, 2015), thereby also extending their use.

There is, however, considerable risk associated with recycled materials. There is firstly incomplete information with regard to the quality and property of these recycled elements. Suppliers are in control of the information they may choose to share or convey and are not always transparent in informing the consumer. More transparency on the part of the producer would very certainly generate a greater degree of trust among European consumers (Milios, 2018). Furthermore, the development and establishment of harmonised standards for recycled or secondary raw materials would potentially help to overcome such barriers (Finnveden et al., 2013).

5.2.3 Stage 3: Waste Management

Waste management is a fundamental element of CE and reducing waste generation is key to achieving the goals of the CE Action Plan.

Today, on average, only 40% of the waste produced by EU households is recycled; however, this figure varies between Member States from 80% in some Member States to only 5% in others. The EC is to develop a long-term perspective on reducing the landfilling of municipal waste end increasing recycling while considering the differences between Member States (European Commission, 2015). The actual quality of the recycling will also need to be improved (i.e., waste collection and sorting instead of incineration). A further obstacle preventing higher recycling rates is the illegal transport of waste both within and outside of the EU, which is often environmentally unsound. Furthermore, promoting the conversion of "waste to energy" of the non-recycled waste is very much aligned with the principle of CE.

5.2.4 Stage 4: From Waste to Resources: Boosting the Market for Secondary Raw Materials and Water Reuse (i.e. Competitiveness and Innovation)

Secondary raw materials refer to the recycled materials that are injected back into the economy. These materials may be shipped just like any primary raw materials.

The procedures of waste management affect directly the quality and quantity of secondary raw materials, therefore increasing and improving the use of these becomes essential. Furthermore, there is considerable uncertainty around the quality of secondary raw materials, and a general lack of trust in them on the part of the consumer. The Commission, therefore, aims to develop quality standards for these raw materials (i.e., presently these standards differ between different Member States), so they will no longer be legally considered to be simply "waste," thereby rekindling consumer trust in such secondary raw materials.

There are several secondary materials that should be given special consideration. The first of these is organic waste material that may be returned to the soils as fertilisers, with the objective of reducing their environmental impacts, while also reducing the dependency on phosphate rocks, which are the basis for mineral-based fertilisers. Water is a valuable commodity experiencing demand pressure. Water-efficiency measures (i.e., as well as the treatment of wastewater) are necessary to relieve pressure on the limited water resources in Europe.

Furthermore, there is also the problem of chemicals. There are several chemicals that have been identified as being harmful to human health and some of these may be present in certain recycling streams (i.e., with the products being sold before the specific restriction on certain chemicals was applied). Therefore, non-toxic streams are encouraged and so is the tracking of chemicals of concern, in order to reduce potential burdens on recyclers, thereby also increasing the trust in secondary raw materials (European Commission, 2015).

5.3 Barriers to the Implementation of CE in Europe

CE is to date an approach that has been managed and driven by practitioners. It has been argued that policymakers and economic stakeholders do not know very much about the basic principles of CE, nor about their impact on the economy (Stahel, 2013; Milios, 2018). At the academic level, literature is only being developed and therefore more extensive research is necessary regarding the concept, as well as its units of analysis. At the academic level, CE academic publications have more than doubled since 2013 (Sacchi et al., 2018).

According to Korhonen et al., there is a perception of there being considerable limitations to the applicability of CE, a perception that could perhaps be overcome by the development of a better definition of the term (Korhonen, et al., 2018). Korhonen et al. (2018) present the following definition: "CE is a sustainable development initiative with the objective of reducing the societal production and consumption systems' linear material and energy throughput flows by applying materials cycles, renewable and cascade-type energy flows to the linear system. CE promotes high-value material cycles alongside more traditional recycling and develops systems approaches to the cooperation of producers, consumers and other societal actors in sustainable development work."

CE in itself is a difficult-to-implement concept, and novel strategies to put CE into action are needed (Kirchherr et al., 2018). Furthermore, implementing CE in the EU needs considerable changes in the actual production and consumption processes and societal behaviour (Kirchherr et al., 2018), such as the adoption and adaptation to novel technologies, changes in market activities, new regulations as well as significant cultural changes.

According to De Jesus and Mendonça (2018), there are four main categories of barriers to the implementation of CE in Europe: (1) "cultural," (2) "regulatory," (3) "market" and (4) "technological." Kirchherr et al. (2018) have taken these four categories and developed a nested model, where the components of one category can determine the components of another. Furthermore, there are interactions between these different components and those that are not CE-compliant may end up in a production or supply chain and then risk provoking the failure of the implementation of the CE in the whole chain and thus obliging the undesired continuance of the existing linear model. Kirchherr et al.'s (2018) CE barrier table can be viewed in Figure 5.1.

From a system's perspective, to be able to successfully transition to and implement CE in Europe, there is the necessity for all stakeholders within the society to participate interdependently and this may only be achieved through the development of suitable collaboration and exchange scenarios (Milios, 2018). This is applicable at the economic, organisational and environmental levels. Furthermore, a difficult scenario for policy development is created by the complex panorama resulting from the multiple possible stakeholder interactions that may affect directly or indirectly part of the value chain (Milios, 2018).

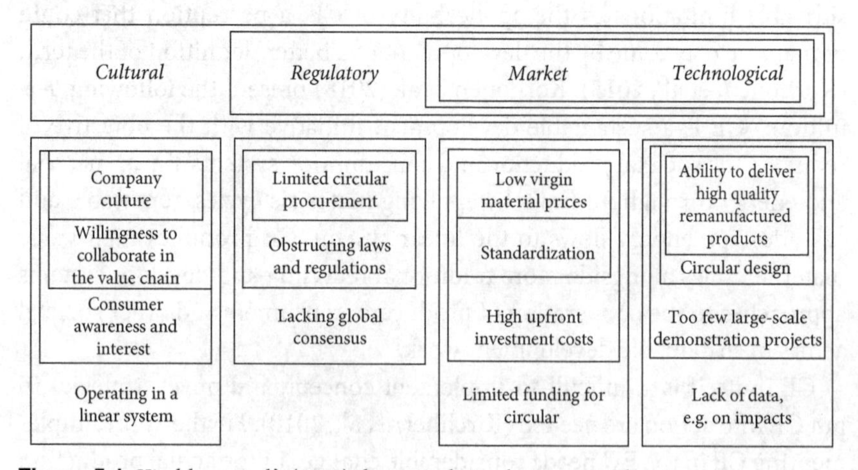

Figure 5.1 Kirchherr et al.'s (2018) theoretical circular economy barriers.

As shown in Figure 5.1, an important potential impediment to implementing CE is the cultural barrier that may be experienced at the consumer or company level. Often at the company level, the adoption of CE may be limited to the environmental or sustainability department, and there is thereby a lack of integration in the possibly varying approaches within a business (Ritzén and Sandström, 2017; Kirchherr et al., 2018), often due to differing perspectives and the absence of communication between different company departments. Moreover, there is also hesitation (i.e. a "hesitant company culture") and a fear of change, part of which may be explained by the uncertainty with respect to market demand for CE. The move towards a CE business model necessitates a change to be implemented throughout the whole of the organisation, and thereby a greater degree of integration and participation/communication between the organisation's different departments (Ritzén and Sandström, 2017). The cultural barrier may also be from the consumer end of the spectrum, and in general, there may be a lack of consumer interest and awareness about CE (Kirchherr et al., 2018). In a study, Ritzén and Sandström (2017)found that there was a need for a deeper understanding of CE and that this lacuna prevented a fluid evolutionary change towards CE.

With regard to the market barriers, primary raw materials are in many cases cheaper than secondary (i.e., or recycled materials). Therefore, why pay more when you can pay less? For example, in the case of plastics, fossil-fuel derived plastics are more economical than biodegradable plastics. Furthermore, setting up a CE project can be costly and often involve elevated upfront investment costs, and may be considered "too costly" to fund (Kirchherr et al., 2018).

With regard to the regulatory frameworks, these often lack specific legal support for CE, and some regulations and laws may actually be obstructive (Kirchherr et al., 2018). There are regulations, for example, that limit the transport of secondary materials across national borders. Such barriers should be removed in order to ensure the smooth implementation of CE models. Governments should consequently step up to keep up with the CE momentum via, for example, subsidies.

Furthermore, at present, although there are promising measures that have been developed by the EU such as eco-labelling and Green Public Procurement, that are of a regulatory nature, they are still not binding, thereby leaving countries to decide on a voluntary basis whether to implement them or not. It may also be argued that these measures are rather general in their nature and do not target, for example, specific materials to become resource-efficient, thereby demonstrating a policy gap (Milios, 2018). Furthermore, the extent of the application of Green Public Procurement has been found to vary very much geographically across the EU territory and not to have been fully implemented. This may be due to the size of the governmental entity undertaking the procurement, its strategic approach to procurement, the actors involved in the procurement process as well as a possible lack of knowledge thereof (Bratt et al., 2013; Guenther et al., 2013; Preuss, 2007; Marron, 2003).

A further barrier is technology or the lack thereof. CE involves the use of secondary or recycled materials. These materials may be very different from primary raw materials, as regards, for example, their chemical composition. Consequently, the right technology must be available to ensure a smooth transition to CE and to guarantee the correct quality. Technological bottlenecks are therefore the main challenge to overcome. A big part of this technological barrier is the design concept, which often is not adequate to promote the development of CE. It is for this reason that the CE Action Plan aims for an improved product design to facilitate repair, upgrade and remanufacture (i.e., through the Ecodesign Directive, Directive 2009/125/EC). This Directive covers all energy-related products.

The last barrier that may be encountered when transitioning towards a CE is being able to acquire a qualified workforce with the right skills to enable this transition. To help overcome this barrier, the EC has a Green Employment Initiative, whose objective is to promote the development of the skills needed to support a green economy.

If we go back to the point about maintaining the value of products, materials and resources in the economy for as long as possible, which is the core of CE (i.e., according to the EU definition of the term presented at the beginning of this chapter), we might think about and question the limits of the circulation of materials in our economy and whether after some time there may be rebound effects (Zink and Meyer, 2017; Milios, 2018). Therefore, the CE's closed-loop economy may have its limits, and this could be when the circular material flow exceeds the corresponding benefits to society (Milios, 2018). The life of these material loops should thus end when the benefits to society are exceeded by the costs of keeping these materials in the economy. Furthermore, there is also the argument that 100% recyclability is not possible, as all materials lose properties after a specific period of time. Also, as the global population is increasing, the demand on materials may be too great and it might not be possible to "close the loop" as regards to the production and consumption cycles of materials (Milios, 2018).

5.4 Funding CE in Europe: Innovation and Investment

In order to enable CE to flourish, it is essential to create the right conditions, which involves the creation of investment opportunities for CE initiatives. Many of these initiatives concern innovation to develop value-added products from waste, which involve new processes, technologies, services and business models (European Commission, 2015). Therefore, innovation and research are key to contributing to competitiveness and modernisation of EU industry.

Horizon 2020 is a funding programme in the EU for research and innovation (European Commission, 2018e). For 2016–2017, the Horizon 2020 programme included a major initiative named, "Industry 2020 in the circular economy," which granted over 650 million euros to projects dealing with CE or industrial competitiveness (European Commission, 2015). A further two funding schemes developed by the EU are that of the Cohesion Policy and the LIFE Programme (i.e., for which CE is a funding priority). The Cohesion Policy aims to promote a more comprehensive application of CE, covering especially rural development and maritime policy (European Commission, 2018f). Established in 1992, the LIFE programme funds principally practical and innovative initiatives dealing with biodiversity and habitat loss, effective resource use and the prevention of climate change (Wysokinska, 2016).

At the private level, funding must be directed towards those opportunities that have been brought about by CE. At the EU level, there is also the European Fund for Strategic Investments, the European Investment Fund and the European Investment Advisory Hub that may also be used to fund these opportunities (European Commission, 2015).

Although there are successful examples of companies that have incorporated CE in their corporate DNA and strategic vision, there is still a need to promote greater investment from both the public and private sectors in order to develop CE initiatives and projects.

There are examples of successful European companies that have based their business on the concept of CE and have consequently been successful. One example of these is the Spanish company "Ecoalf," which produces clothes and shoes from recycled plastics (i.e., the latter of which have been made from the plastic gathered from oceans in Spain and Thailand). Specifically, in Spain, Ecoalf works with fishermen to gather plastics from the Mediterranean Sea, to consequently extract polymers and make new fibres and clothes. Since its establishment in 2012, the company has been successful in getting on its feet a business based on sustainability (i.e., CE) through innovation and the development of a successful Unique Selling Point (USP). Since 2012, Ecoalf has opened up two stores, one in Madrid (Spain) and another in Berlin (Germany). Innovation, however, comes at its price, and Ecoalf clothes and shoes may unfortunately not be affordable by everyone.

On a more global level, there are several other companies that have included CE in their corporate DNA and consequent company commercial image. For example, H&M has been one of the first companies to sign the Fashion Industry Charter for Climate Action, which was established during the last Climate Change Conference (COP24), held on December 2018 in Katowice, Poland (H&M, 2018). The Company has set itself clear sustainability goals, such as, for example, using renewable energy in all its operations, decreasing its energy use, and committing itself to a climate-neutral supply chain by 2030. IKEA is another exemplary company that has incorporated sustainability in its corporate DNA. Its logo "People and Planet Positive" (Ikea, 2018) reflects this commitment.

5.5 CE Priority Areas

There are several priority areas that face certain challenges with respects to CE that were identified in the Circular Economy Action Plan, such as the plastics problem, food waste, critical raw materials and biomass. In this chapter, however, most of the discussion on CE priority areas will be based on the plastics problem and the funding issue for CE projects in Europe.

5.5.1 A New Plastics CE

Plastics have been identified as crucial throughout the whole of the value chain. The main objective here is that all plastic packaging should be recyclable by 2030. The plastics issue has also identified as a key priority of the EU Action Plan for CE, an initiative addressing the challenges that affect the whole of the value chain (European Commission, 2018a).

Over the past few years, "plastic" has become a commonly used material in our daily lives and economy. However, if plastics are not disposed of properly, their effects on the environment can be profound since the products can last for long periods and degrade into what are known as microplastics. The problem of plastic waste is aggravated every year and is exacerbated by the growing consumption of "single-use" plastics (European Commission, 2018a). With regard to environmental contamination, an example of the latter is the problem of ocean contamination, where millions of tonnes of plastic litter end up in the seas and oceans year after year, affecting wildlife, and at the end of the day also affecting human lives.

Plastic has, over the last few decades, become a widely used material and has helped our society to become more effective in areas such as the automotive, construction and food sectors. It has been of particular use in the packaging of all types of materials. In the food sector, for example, it has helped to ensure food safety, and consequently reduce food waste.

The EU Action Plan for CE aims to create a new plastics economy, which is based on the principles of reuse, repair and recycling, thereby taking on the present challenges and turning them into opportunities. The idea is also to develop and promote more sustainable materials and turn Europe into the leader for this transition, boosting innovation and consequently enhancing prosperity in Europe through the concept of sustainability. This new plastics economy is "modern, low-carbon, resource-efficient and energy-efficient" (European Commission, 2018a).

5.5.2 The Plastic Problem at Present

The importance and prominence of the plastic economy in Europe have increased considerably over the past several decades. It is estimated that the global production of plastics has increased 20-fold since the 1960s and is expected to double again over the next 20 years (European

Commission, 2018a). About 59% of the plastic generation in the EU derives from packaging, with the rest originating from the sectors of electrical and electronic equipment production, construction, automotive and agriculture. Furthermore, the plastic industry in Europe employs about 1.5 million people (European Commission, 2018). As the role of plastics in our society has become more pronounced, so has grown the plastic waste that is generated annually on a global basis. In the EU alone, it is estimated that around 25.8 million tonnes of plastic waste is generated every year (European Commission, 2018). Most of this waste is the result of our current linear economy, which is based on the traditional economy of model based on "take-make-consume-throw away approach of resources" (EIONET, 2018).

Although CE is now being promoted by the EC, at present most of the waste that is collected is not recycled. All this plastic waste is causing considerable environmental problems and distress, accounting for 31%–39% of all waste generated (European Commission, 2018a). Furthermore, the production and incineration of plastic accounts for the production of an average of approximately 400 million tonnes of CO_2 every year, thereby greatly contributing to the accumulation of greenhouse gases in the atmosphere.

The environmental problems due to plastics do not stop there. It is estimated that 1.5%–4% of the global plastic production ends up in the seas and oceans every year (Jambeck et al., 2015), and this, in turn, is estimated to account for 80% of all the marine litter (European Commission, 2018a). In Europe, the stats show that about 150,000–500,000 tonnes of plastic end up in European waters every year (European Commission, 2018a). These figures may seem small when compared to the global states; however, marine currents are able to transport plastics over very long distances, thereby making the plastic issue a global problem and not a localised issue. The taking of measures to curb the problem in one part of the world is bound to affect another area with the same problem in some other part of the globe, thereby highlighting the importance and essentiality of the EU Action Plan for CE.

Referring more specifically to the European geography, the accumulation of plastics tends to be most notable in vulnerable marine areas such as the Mediterranean Sea. Adverse effects of plastic in the oceans include the entanglement and ingestion of the plastic materials by wildlife, causing the deaths of thousands of fish, seabirds, sea turtles, seals and other marine animals. According to the Center for Biological Diversity (CBD), there are over 700 species that eat or get caught in plastic litter (Center for Biological Diversity, 2018). Additionally, the disposal of these massive quantities of plastic waste in the European marine environment brings further economic damage to such activities as tourism, fisheries and shipping (European Commission, 2018a).

A further problem that is now evident in European waters is that of the microplastics. According to the European Chemical Agency (ECHA), microplastics are "very small particles of plastic material (typically smaller than 5 mm)" (ECHA, 2018). These may be dumped directly as microplastics into the oceans or be the result of the degradation of larger pieces of plastic. Microplastics enter the human food chain through their initial ingestion by fish. Furthermore, microplastics have been found in the air and drinking water; however, the exact effects of these on human health have yet to be determined. Microplastics are also considered to be a cross-border issue that needs to be tackled globally.

5.5.3 *Transitioning Towards a New Plastics Economy*

There is presently a caveat with regards to the market potential from bringing about a CE for plastics. According to the Ellen MacArthur Foundation, 95% of the value of plastic packaging material is lost to the economy after very short first-use life cycles (Ellen MacArthur Foundation, 2016) Furthermore, the demand for recycled plastic materials is only 6% of the plastic demand in Europe, which demonstrates that there is a clear market potential for the use of recycled plastic in today's society, which could bring considerable benefits, both economic and non-economic. A "plastics CE" is therefore definitely the way forward.

However, with so much potentiality and capacity, why isn't this "plastic CE" more prominent or common today. What are the problems that are preventing this from becoming a reality? We could perhaps firstly argue that the demand in our society for these products is not there. This could be due to three main reasons: (1) that there may be a general lack of awareness in European society with regard to the plastics issue and the negative impacts that it can have on the environment and on their own health, (2) there may be a considerable level of uncertainty in regard to the quality of recycled products, are they equivalent in quality to the products people are now used to and (3) some of the products that have been partly or fully produced using recycled plastics can be more expensive than normal store items. Most of us are very much aware of how much we spend and thereby investing a bit more money in a coat, for example, that has been produced from recycled materials, may be less desirable when the home budget or economy is tight. The recent economic crisis that Europe experienced in 2008–2012 has not aided this transition towards a circular plastics economy in any way. With tight budgets, it was very difficult for Europeans to spend additional money on recycled materials that cost more.

The elevated levels of uncertainty from the demand side to support a circular plastics economy is equally present at the producer end of the spectrum. The 6% statistic of total recycled plastic demand in Europe (European Commission, 2018a) may be justified by the elevated degree of uncertainty with regards to the potential profitability to be obtained by a company that uses recycled plastics in the development of its products. Therefore, measures are needed on the part of the European Union to increase trust levels among European companies and get them to use in the future-recycled plastic materials for their production. Furthermore, there is also the fear that companies that have those recycled plastics may not meet their needs for continuous volume supplies of materials that need continuous quality specifications (European Commission, 2018a), hence the requirement for the standardisation of specific equipment components in order to facilitate the process of transition towards Europe's new plastics economy.

From the EU's perspective, there is a present need to look ahead and set out clearly what a future "circular" plastics economy vision will entail for both the short and long term (European Commission, 2018a). This will require the determination of concrete measures on the part of the EU to make this vision a reality. The correct stakeholder identification, planning and participation (i.e., which will be later described in more detail) will also be essential to ensure the correct implementation of this strategy and bring about positive change.

5.5.4 Europe's New Plastics Economy

The vision for Europe's new plastics economy prioritises the importance of reusing, repairing and recycling, which will through time create more jobs in Europe and generate growth while helping to curb greenhouse emissions (European Commission, 2018a). The benefits are therefore both economic and environmental; however, these will only be achieved if all key players participate in the transition; from producers to consumers. Key players may also include the civil society, the scientific community, businesses and local authorities as well as both regional and national governments.

Very important here is the participation of the European citizens, as consumers. Ensuring that civil society supports sustainability and sustainable initiatives is indispensable to facilitate and create the fertile ground for this transition towards a sustainable plastics economy. Citizens, therefore, as consumers, should be made aware of the need to avoid waste, as well as on the main benefits that will be obtained from this transition towards a new plastics economy.

The priorities of this new plastics economy will include the following according to the EC (European Commission, 2018a: (1) ensure the durability of plastics and plastic-containing products by 2030. This includes plastic packaging that should be recyclable or recycled in a cost-effective manner; (2) design and production processes should develop products with higher recycling rates. This especially applies to packaging, which accounts for 60% of post-consumer plastic waste. It has been shown that design improvements could reduce by half, the cost of recycling plastic packaging waste (Ellen MacArthur Foundation, 2017). The 2030 objective is for all plastics packaging on the EU market to be reusable or easily recycled. There are also additional guidelines to reward the most sustainable designs. Moreover, there is a present need for standardisation with regards to the materials making up the different products. Such is the case for the sector of electrical appliances and electronic goods, where the standardisation of equipment will facilitate its dismantling, reuse and recycling; (3) recycling capacity should be modernised and increased, thereby creating thousands of new jobs in Europe; (4) there should be greater integration within the plastics value chain, and any obstacles within the chain should be removed; (5) although difficult, it is essential for there should be the successful establishment of a market for recycled plastics; (6) increasing plastics recycling will reduce the dependence on fossil fuels, as well as CO_2 emissions (in line with the Paris agreement); (7) using alternative and innovative materials as a base to produce plastics (i.e., this, in turn, will lead to further employment opportunities, as well as opportunities for growth) and (8) having Europe continue its leadership with regards to the recycling of equipment and technologies.

5.6 The CE Monitoring Framework

The CE Monitoring Framework was developed in December 2018 to monitor how CE was developing over time in Europe, and thereby determining how much success the EU Circular Action Plan has had so far. Furthermore, the development of this framework will help to set up long-term priorities of a CE in Europe.

A set of 10 indicators have been developed to monitor the CE in Europe as part of this monitoring framework. The framework is not limited to certain materials, sectors or specific geographical locations. Furthermore, it is impossible to develop one single indicator to account for "circularity" (European Commission, 2018b). These 10 indicators are categorised into four stages of CE (i.e., from the point that the materials are extracted from their source to that when they become waste) as listed below: (1) production and consumption (EU self-sufficiency for raw materials, green public procurement, waste generation, food waste); (2) waste management (overall recycling rates, recycling rates for specific streams); (3) secondary raw materials (contribution of recycled materials to raw material demand, trade in recyclable raw materials) and (4) competitiveness and innovation (private investments, jobs and gross value added, patents).

5.7 Initial EU CE Findings

From the establishment of the EU CE Action Plan in December 2015, three years on some first findings have been compiled as regards the strategy's success. This will form the basis for the improvement of these indicators and the monitoring framework, by the setting up of baselines from which potential benchmarking may be carried out. The Framework in itself is a tool that is especially useful to follow main trends in the transition towards a CE, and thereby determine whether the implemented measures and the stakeholder engagement have been to date effective enough to enable the identification of best practices in Member States (European Commission, 2018b).

A few of the first initial findings are listed below concerning the 10 indicators (i.e., all of these findings derive from the Communication from the EC regarding the Monitoring Framework for CE published in December 2018):

- *Production and consumption*: The EU may be considered to be in general terms "self-sufficient", that is except for some critical raw materials (which are essential for the EU to attain a sustainable, low-carbon economy). Data are still pending regarding the indicator of green public procurement. Municipal waste generation per capita has dropped by 8% from 2006 to 2016 (i.e., however, large differences still exist between Member States). The reduction of food waste is applicable to all of the value chain. EU food waste was found to have decreased by around 7% from 2012 to 2014 (European Commission, 2018f).
- *Waste management:* Recycling rates for municipal waste have increased from 37% to 46%. Increases were also found to be a reality for packaging waste, biowaste, and construction and demolition waste. Recycling rates were found to vary between Member States for waste electrical and electronic equipment.
- *Secondary raw materials:* On average, recycled materials only make up circa 10% of the EU demand for materials. As regards to the indicator of trade of recyclable raw materials, it is shown that the EU exports several recyclable waste streams such as those of plastics, paper and cardboard, copper, aluminium, nickel and other precious metals.
- *Competitiveness and innovation*: Transitioning to a CE increases investments, innovation and value-added jobs (European Commission, 2018b). In 2014, 15 billion euros were invested in CE-associated economic sectors and 3.9 million jobs created. Moreover, there are a number of EU funding programmes that support the CE transition, such as the European Fund for Strategic Investments, the European Structural and Investment Funds, Horizon 2020 and the LIFE Programme. Finally, in January 2017, the CE Finance Support Platform was inaugurated (Business Review, 2018). Between 2000 and 2013, patents for recycling and secondary raw materials increased by 35%.

Conclusion

The foundations for a successful CE Action Plan have been established by the EU. Clearly, transitioning to a CE has its multiple benefits; however, there is still a long way to go. The results to date have been positive, but there are still many stakeholders to convince and involve. It will only be when there is a quorum and that perspectives are aligned with regards to the implementation of circular actions that the CE Action Plan will really be a success.

References

Bourguignon, D. (2016) *Closing the Loop; New Circular Economy Package.* European Parliament, available at: http://www.europarl.europa.eu/RegData/etudes/BRIE/2016/573899/EPRS_BRI(2016)573899_EN.pdf

Bratt, C., Hallstedt, S., Robert, K.H. and Oldmark, J. (2013) Assessment of criteria development for public procurement from a strategic sustainability perspective. *Journal of Cleaner Production,* 52, 309–316.

Brings Jacobsen, N. (2006) Industrial symbiosis in Kalunborg, Denmark: A quantitative assessment of economic and environmental aspects. *Journal of Industrial Ecology,* 10(1–2), 239–255.

Business Review (2018) *CE Finance Platform.* Business Europe Website, available at: http://www.circulary.eu/timelines/circular-economy-finance-support-platform-separator-ongoingv2/

Center for Biological Diversity (2018) Ocean Plastic Pollution, available at: https://www.biologicaldiversity.org/campaigns/ocean_plastics/

Costea-Dunarintu, A. (2016) The circular economy in the European Union. *Knowledge Horizons – Economics,* 8(1), 148–150.

De Jesus, A. and Mendonça, S. (2018) Lost in transition? Drivers and barriers to the eco-innovation road to the circular economy. *Ecological Economy,* 145, 75–89.

Ellen MacArthur Foundation (2016) *The New Plastics Economy; Rethinking the Future of Plastics.* World Economic Forum, Ellen MacArthur Foundation and McKinsey & Company, available at: https://www.ellenmacarthurfoundation.org/assets/downloads/EllenMacArthurFoundation_TheNewPlasticsEconomy_Pages.pdf

Ellen MacArthur Foundation (2017) *The New Plastics Economy: Catalysing Action,* available at: https://www.ellenmacarthurfoundation.org/assets/downloads/New-Plastics-Economy_Catalysing-Action_13-1-17.pdf

EIONET (2018) Linear Economy. *European Environment Information and Observation Network,* available at: https://www.eionet.europa.eu/gemet/en/concept/15216

Enkvist, P. A. and Klevnäs, P. (2018) *The Circular Economy a Powerful Force for Climate Mitigation.* Material Economics: Stockholm, available at: https://www.sitra.fi/en/publications/circular-economy-powerful-force-climate-mitigation/

EU Directive (2009) *Establishing a framework for the setting of ecodesign requirements for energy related products* – European Union Directive 2009/125/EC of the European Parliament and the Council. Official Journal of the European Union.

European Commission (2015) *Closing the loop – an EU action plan for the Circular Economy. COM. 614 final.*

European Commission (2016a) *Ecodesign Working Plan 2016–2019. COM. 773 final.*

European Commission (2016b) Growing the circular economy, Building a stronger, greener and more sustainable future. Panorama 50.

European Commission (2017a) *Invertir en una industria inteligente, innovadora y sostenible Estrategia renovada de política industrial de la UE. COM. 479 final.*

European Commission (2017b) Happy Birthday EU Ecolabel! Available at: https://ec.europa.eu/environment/efe/news/happy-birthday-eu-ecolabel-2017-06-21_en

European Chemicals Agency (2018) Microplastics. Available at: https://echa.europa.eu/hot-topics/microplastics

European Commission (2018a) *A European Strategy for Plastics in a Circular Economy. COM. 28 final.*

European Commission (2018b) *On a monitoring framework for Circular Economy. COM 29 final.*

European Commission (2018c) *Ecodesign*, available at: http://ec.europa.eu/growth/industry/sustainability/ecodesign_en

European Commission Green Public (2018d) *Green Public Procurement,* available at: http://ec.europa.eu/environment/gpp/index_en.htm

European Commission (2018e) *What is Horizon 2020?*, available at: https://ec.europa.eu/programmes/horizon2020/en/what-horizon-2020

European Commission (2018f) Cohesion Fund, available at: ://ec.europa.eu/regional_policy/en/funding/cohesion-fund/

European Commission (2018g) *EU Actions against Food Waste,* available at: https://ec.europa.eu/food/safety/food_waste/eu_actions_en

Finnveden, G. et al. (2013) Policy instruments towards a sustainable waste management. *Sustainability,* 5(3), 841–881.

Ghisellini, P., Cialani, C. and Ulgiati, S. (2016) A review on circular economy: The expected transition to a balanced interplay of environmental and economic systems. *Journal of Cleaner Production*, 114, 11–32.

Guenther, E., Hueske, A.K., Stechemesser, K. and Buscher, L. (2013) The "Why Not" – perspective of green purchasing: A multilevel case study analysis. *Journal of Change Management,* 13(4), 407–423.

Günther, E. and Scheibe, L. (2006) The hurdle analysis. A self-evaluation tool for municipalities to identify, analyse and overcome hurdles to green procurement. *Corporate Social Responsibility and Environmental Management,* 13(2), 61–77.

H&M (2018) *H&M Group joins the fashion industry charter for climate action.* H&M Group website, available at: https://hmgroup.com/media/news/general-news-2018/hm-group-joins-UNFCCC.html

Holton, J., Glass, J. and Price, A.D.F. (2010) Managing for sustainability: Findings from four company case studies in the UK precast concrete industry. *Journal of Cleaner Production,* 18, 152–160.

Ikea (2018) Ikea Sustainability Strategy People and Planet Positive. *Inter IKEA Systems BV,* available at: https://www.ikea.com/ms/en_US/pdf/people_planet_positive/IKEA_Sustainability_Strategy_People_Planet_Positive_v3.pdf

Iveroth, S.P., Johansson, S. and Brandt, N. (2013) Implications of systems integration at the urban level: The case of Hammarby Sjöstad in Stockholm, Sweden. *Journal of Cleaner Production,* 59, 716–726.

Jambeck, J.R., et al. (2015) Plastic waste inputs from land into the Ocean. *Science,* 347, 768–771.

Kirchherr, J. et al. (2018) Barriers to circular economy: Evidence from the European Union. *Ecological Economics,* 150, 264–272.

Korhonen, J., Nuur, C., Feldmann, A. and Echetu Birkie, S. (2018) *Journal of Clean Production,* 175, 544–552.

MacArthur, E. (2015) *Towards the Circular Economy. Economic and Business Rational for an Accelerated Transition*. Ellen MacArthur Foundation: Cowes.

Maitre-Ekern, E. and Dalhammar, C. (2016) Regulating planned obsolescence: A review of legal approaches to increase product durability and reparability in Europe. *Review European Community and International Environmental Law,* 25(3), 378–394.

Marron, D. (2003) Greener public purchasing as an environmental policy instrument. *OECD Journal on Budgeting,* 3(4), 71–105.

Milios, L. (2018) Advancing to a circular economy: Three essential ingredients for a comprehensive policy mix. *Sustainable Science,* 13, 861–878.

Preuss, L. (2007) Buying into our future: Sustainability initiatives in local government procurement. *Business Strategy and the Environment,* 16(5), 354–365.

Ritzén, S. and Sandström, G.Ö. (2017) Barriers to circular economy – integration of perspectives and domains. *Procedia CIRP,* 64, 7–12.

Sacchi Homrich, A., Galvao, G., Gamboa Abadia, L. and Carvalho, M.M. (2018) *Journal of Cleaner Production,* 175, 525–543.

Stahel, W.R. (2013) Policy for material efficiency – sustainable taxation as a departure from the throwaway society. *Philosophical Transactions of the Royal Society of London A: Mathematical, Physical and Engineering Sciences,* 317(1986), 1–17.

The Paris Agreement (2019) available at: https://unfccc.int/process-and-meetings/the-paris-agreement/the-paris-agreement

Van Berkel, R., Fujita, T., Shisuka, H. and Fujii, M. (2009) Quantitative assessment of urban and industrial symbiosis in Kawasaki, Japan. *Environmental Science and Technology,* 43, 1271–1281.

Wysokinska, Z. (2016) The "new" environmental policy of the European Union: A path To development of a circular economy and mitigation of the negative effects of climate change. *Comparative Economic Research,* 19(2), 57–73.

Yong, R. (2007) The circular economy in China. *Journal of Materials Cycles Waste Management,* 9, 121–129.

Yu, C., Davis, C. and Dijkema, G.P.J. (2013) Understanding the evolution of industrial symbiosis research. *Journal of Industrial Ecology,* 18, 280–293.

Zink, T. and Meyer, R. (2017). Circular economy rebound. *Journal of Industrial Ecology,* 21(3), 593–602.

CHAPTER

6

The Energy Union: The EU at Work

María Isabel Nieto

If you want to go fast, go alone; if you want to go far, go with others.

–African proverb

Access to energy is today an indispensable condition for our subsistence and the maintenance of our way of life. Lighting, transport, housing, heating, industrial production and operation of household appliances, among others, require a constant energy supply at an affordable price which allows the competitiveness of European companies that make use of this energy for the generation of products and services to be maintained. According to data from the European Commission (EC), the European Union (EU) consumes a fifth of the world's energy, which means an annual cost of more than 400,000 million Euros (European Commission, 2017), despite the fact that the internal reserves are scarce. One of our greatest challenges is the promotion of the use of renewable resources such as wind and solar power, and to favor research on how to improve their performance and utilization (the latter is very difficult due to the absence of a single energy market and the lack of interconnections between national networks) and how to reduce their environmental impact while investing in research for the discovery and use of further new sources of clean energy. The EU has a clear commitment for taking steps from the dependence on fossil fuels to the employment of clean energy and is carrying out measures in this direction, although still very insufficient.

In the Energy and Climate Change Framework, short-term objectives for 2020, medium-term for 2030 and long-term for 2050 are on the European horizon (Figure 6.1).

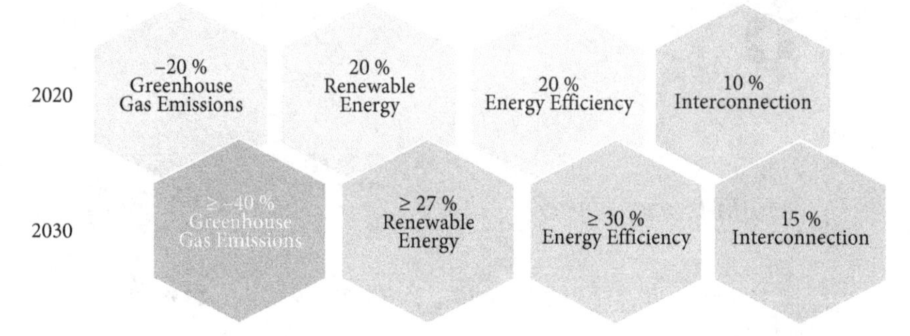

Figure 6.1 2030 Framework for Climate and Energy – Agreed Headline Targets.
Source: European Commission, Third Report on the State of the Energy Union, COM(2017) 688 final.

The environmental objectives of the EU clearly orientated toward the decarbonization of its Member States and in line with the commitments made with the entry in force of the Paris Agreement, aim to achieve an integrated, interconnected, competitive and diversified market in order to reduce dependence on external energy sources and steadily undergo a transition toward a clean and sustainable economy. However, the increase in energy demand, the scarce diversification of supply sources, the fluctuations in prices, the high-energy dependence of some member countries (the EU is the largest importer of gas at a global level), the shortcomings of infrastructures and the limited interconnection of national networks and the risks that these shortcomings can pose for the continuity of supply, along with the expansion of the still small share of renewable energies, are the key challenges that the Union faces in achieving these energy objectives. The EU's main obstacle to progress in renewable energies, as in many other issues, is that a large part of the responsibility lies in the commitment of its Member States, and in the firm adoption of an approach not so focused on internal markets as on a strategic thinking that benefits the Union globally. Only then the competitiveness of the EU and its global technological leadership will result in being strengthened.

Energy policy is, therefore, a shared competence between the Union and its Member States, established according to the ordinary legislative procedure, which is moving very slowly toward a common energy policy (except measures of a fiscal nature or taxes on energy that are established in accordance with a special unanimous legislative procedure): "If the EU seeks to legislate in a way that significantly affects these Member State rights, then a special legislative procedure must be followed which requires an initial decision by unanimity in Council" (European Parliamentary Research Service, 2018). Although it is up to the EU to guarantee continuity of the supply (focusing mainly on the gas and electricity markets) in accordance with the Treaties, the Member States are responsible for determining the general structure of their energy supply, the choice of the different energy sources, better known as the "energy mix", along with the conditions of exploitation of its energy resources.

The legal basis is set out in Article 194 of the Treaty of Functioning of the European Union (TFEU)(OJEU, 2012). Its content, introduced by the Treaty of Lisbon, states that "in the context of the establishment and functioning of the internal market and with regard for the need to preserve and improve the environment, Union policy on energy shall aim, in a spirit of solidarity between Member States, to a) ensure the functioning of the energy market; b) ensure security of energy supply in the Union; c) promote energy efficiency and energy saving and the development of new and renewable forms of energy; and d) promote the interconnection of energy networks". Further articles concerning this issue are the specific provisions relating to; security of supply (Article 122 TFEU), energy networks (Articles 170–172 TFEU), coal (OJEU, 2016), nuclear energy and other provisions associated with the internal energy market (Article 114 TFEU) and the external energy policy (Articles 216–218 of the TFEU).

In February 2015, in order to respond to these objectives, the EC led by the President Jean Claude Juncker announced its energy strategy, focusing on five dimensions or key areas (Frédéric Gouardères, 2018), which collects and completes the objectives of the Third Report on the State of Energy of the Union:

- Ensure security/continuity of energy supply in the Union
- Promote energy efficiency and energy saving

- Promote the development of new and renewable forms of energy to better align and integrate climate change goals into the new market design
- Promote research, innovation and competitiveness[a]

European public opinion has great expectations regarding these policies. In particular, and concerning the political priorities of the EU (European Commission, 2018), 74% of Europeans are in favor of a common European energy policy, whereas for example, as shown in Figure 6.2, Spain exceeds this percentage with 85% of Spaniards being in favor of such political commitment.

Along the same lines, and as shown in Figure 6.3, the average European citizen agrees that in a European Energy Union maximum priority should be given to the development of renewable energies, protecting the environment, guaranteeing reasonable energy prices for consumers and fighting against global warming.

Likewise, the support of EU citizens for greater EU participation in the field of energy supply and security policy increased from 53% in 2016 to 65% in 2018, which represents an increase of 12 percentage points in the expectations of citizens, the highest among all addressed policy areas, with the strongest support at the individual level in Cyprus and Spain, whereas the weakest support came from the Czech Republic and Austria.

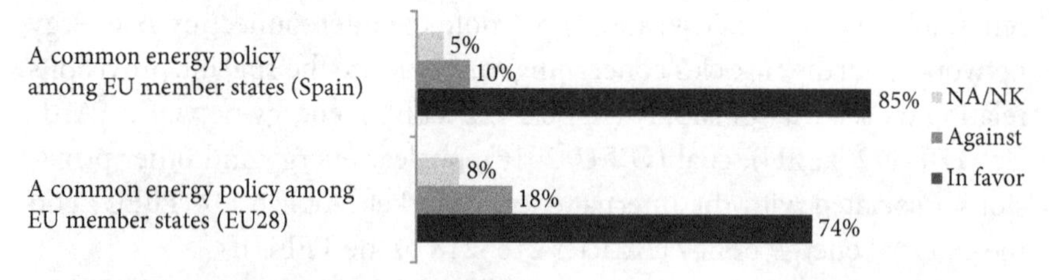

Figure 6.2 Spanish Public Opinion about the need of a Common Energy Policy.
Source: Standard Eurobarometer 90, Autumn 2018.

[a]The main achievements can be found briefly summarized in European Commission (2019).

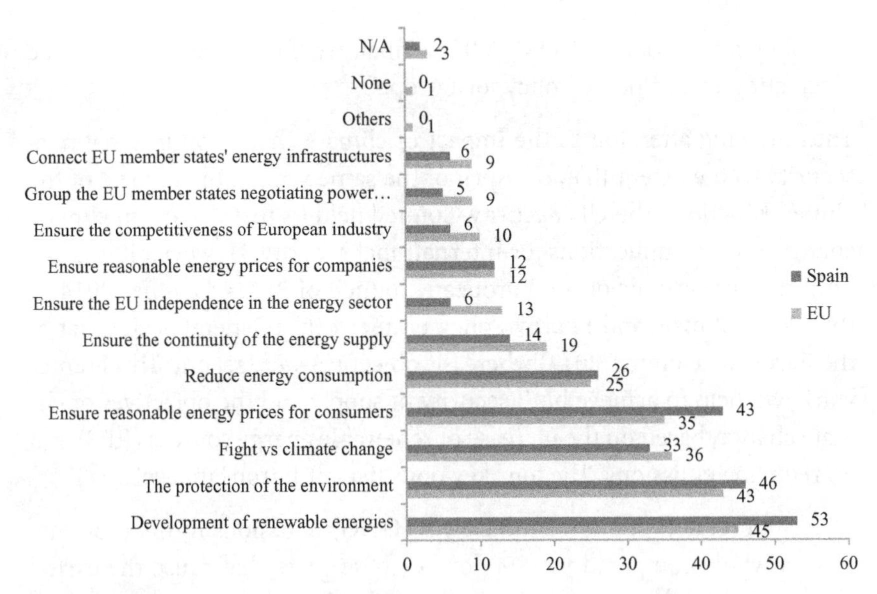

Figure 6.3 Spanish Public Opinion about the priority of the objective of a Common Energy Policy.
Source: Standard Eurobarometer 90, Autumn 2018.

6.1 Current Political Framework of the Energy Policy

The European Council held in Brussels on 8–9 March 2007, drew up an agreement for an integrated climate and energy policy:

- Underline the need to limit the increase of the overall global annual mean surface temperature increase, which "should not exceed 2°C above pre-industrial levels" (Presidency of the European Union, 2007).
- Draw attention to the need for international climate protection and the commitment "to collectively reducing their greenhouse gas emissions by 60 to 80% by 2050 compared to 1990 levels", where it advocates the implementation of an integrated energy policy, which guarantees the security of supply and improves the competitiveness of the companies.
- Adopt a comprehensive energy Action Plan for the period 2007–2009 (Annex I), based on the Commission's Communication "An Energy

Policy for Europe" – COM(2007) 1 final, which is "a milestone in the creation of an Energy Policy for Europe".

Thus drawing attention to the impact of climate change on international security to the extent that, on April of the same year, at the request of the United Kingdom, the UN Security Council held its first debate on climate change and its implications for international security. However, it was not until the conclusions of the European Council of 23–24 October 2014 in the "2030 Climate and Energy Policy Framework" (General Secretariat of the European Council, 2014) where the objectives were set out. This Framework will help to achieve both security of supply and the objectives of climate change, based on the increase of renewable energy sources (RES) and on reducing emissions. The four key objectives at European level are:

- A 40% reduction in greenhouse gas (GHG) emissions in the economy as a whole compared to 1990 levels (these gases that cause the earth's atmosphere to retain the sun's heat are mainly carbon dioxide, methane and nitrous oxide).
- An increase of 27% in the share of renewable energies in energy consumption. The Council itself specified that the objective for renewable energies is binding at EU level and must be met through the contributions of the Member States.
- An increase of 27% in energy efficiency, with a view to further reach 32.5% along with the development of electricity interconnections of 15%. By 2020, it is intended to reach a "minimum target of 10% of existing electricity interconnections, as a matter of urgency, and no later than 2020 at least for Member States which have not yet attained a minimum level of integration in the internal energy market, such as the Baltic States, Portugal and Spain, and for Member States which constitute their main point of access to the internal energy market". Special attention will be paid to the more remote and/or less well-connected parts of the single market such as Malta, Cyprus and Greece. These efforts are in line with the European Energy Security Strategy.

The EU has two financial instruments par excellence aimed at combating GHG emissions:

- On the one hand, the EU Emissions Trading System (ETS) is the cornerstone of the EU's policy to tackle climate change and its key tool for cost-effectively reducing GHG emissions. It is the world's first major carbon market, representing more than three-quarters of the international market of carbon emissions (European Commission, 2019a). It is expected that in 2020, emissions produced by the sectors subject to the regime — carbon dioxide produced in the generation of heat and electricity, sectors of intensive energy consumption including oil refineries, steel mills and production plants for iron, aluminum and other metals, cement, lime, glass, ceramics, paper pulp, cardboard, acids and raw organic chemicals and commercial aviation, as well as nitrous oxide and perfluorocarbons from aluminum production – will be 21% less than in 2005, and in 2030 will be 43% lower. Participation in the ETS is mandatory for companies included in the mentioned sectors with limited exceptions.

- On the other hand, the legislation of shared efforts establishes the annual emission targets for GHG, binding for all 28 Member States for the periods 2013–2020 and 2021–2030. These are objectives referring to sectors not included in the ETS: transport, buildings, waste and agriculture. Thus, for instance, the annual emission targets for 2020 range from a 20% reduction in emissions for the richest states such as Denmark, Ireland and Luxembourg, to an increase of 20% in the poorest states such as Bulgaria, or 19% in Romania. Detailed data per country can be found in Annex II of Decision No 406/2009 / EC (OJEU, 2009).

The emission reduction targets for 2030 for all Member States range from 0% to 40% taking the 2005 levels as a reference. "This Regulation applies to the greenhouse gas emissions from Intergovernmental Panel on Climate Change (IPCC) source categories of energy, industrial processes and product use, agriculture and waste as determined pursuant to Regulation (EU) No 525/2013, excluding greenhouse gas emissions from the activities listed in Annex I to Directive 2003/87/EC" (OJEU, 2018).

The other financial instrument that should be referred to is the *"New Entrants Reserve NER 300"*, a program that finances innovative projects in the field of RES and CO_2 Geological Capture and Storage (CCS) in a way that is respectful to the environment. The program is intended to support a wide range of CCS technologies, namely pre-combustion, post-combustion, oxyfuel and industrial applications, as well as RES technologies, namely bioenergy, concentrated solar power, photovoltaic, geothermal, wind, ocean, hydropower and smart grids (European Commission, 2019b). Two calls for proposals to fund projects in these areas were announced, one in 2012 and another in 2014. The projects that received funding will come into operation in 2021 and no new call for proposals is planned. The EC is focusing on a new *Innovation Fund Programme.* It should be noted that the Commission decided to reinvest the unspent funds of the first round, around 488 million Euros through two other financial instruments already existing in the Union: *the InnovFin Energy Demo Projects (EDP) and CEF Debt Instrument.*

Among the most important legal instruments, the "Clean Energy for all Europeans" COM(2016) 860 (OJEU, 2016) stands out (Table 6.1). This aims to make the EU competitive in these moments of transition toward clean energy. The package covers eight legislative proposals in different areas: governance, the electricity market (through the Electricity Directive, the Regulation on Electricity and the Regulation on Risk Preparedness), energy efficiency, energy-efficient buildings, renewable energies and regulations applicable to the Agency for the Cooperation of Energy Regulators (ACER) which, along with other functions, supervises the wholesale energy markets and warns possible abuses of a dominant position. The EC itself, in a press release from 2018 stated that these proposals are "designed to show that the clean energy transition is the growth sector of the future — that's where the smart money is. Clean energies in 2015 attracted global investment of over 300 billion Euros. The EU is well-placed to use our research, development and innovation policies to turn this transition into a concrete industrial opportunity. By mobilizing up to 177 billion Euros of public and private investment per year from 2021, this package can generate up to 1% increase in GDP over the next decade and create 900,000 new jobs" (European Commission, 2016).

The *Regulation (EU) 2018/1999 of the European Parliament and of the Council of 11 December 2018 on the Governance of the Energy Union and Climate Action* (OJEU, 2018a) also known as the Governance Mechanism should also be mentioned. This legal act implies that a combination of EU initiatives with integrated national energy and climate plans, stipulated for a period of 10 years, will be necessary to achieve the specific objectives of the Energy Union. The plans that will establish national objectives for each of the five dimensions of the Energy Union must be coherent with the

Table 6.1 Clean energy for all Europeans package – state of play (27 March 2019).

	European Commission Proposal	**EU Inter-institutional** Negotiations	**European Parliament** Adoption	**Council** Adoption	**Official Journal** Publication
Energy Performance in Buildings	30/11/2016	Political Agreement	17/04/2018	14/05/2018	19/06/2018 - Directive (EU) 2018/844
Renewable Energy	30/11/2016	Political Agreement	13/11/2018	04/12/2008	21/12/2018 - Directive (EU) 2018/2001
Energy Efficiency	30/11/2016	Political Agreement	13/11/2018	04/12/2018	21/12/2018 - Directive (EU) 2018/2002
Governance of the Energy Union	30/11/2016	Political Agreement	13/11/2018	04/12/2018	21/12/2018 - Regulation (EU) 2018/1999
Electricity Regulation	30/11/2016	Political Agreement	26/03/2019	**Scheduled in May 2019**	-
Electricity Directive	30/11/2016	Political Agreement	26/03/2019	**Scheduled in May 2019**	-
Risk Preparedness	30/11/2016	Political Agreement	26/03/2019	**Scheduled in May 2019**	-
ACER	30/11/2016	Political Agreement	26/03/2019	**Scheduled in May 2019**	-

Source: European Commission, https://ec.europa.eu/energy/en/topics/energy-strategy-and-energy-union/clean-energy-all-europeans, consulted in April 2019.

United Nations sustainable development objectives. Among other things, they must evaluate the number of households in situations of energy poverty, and multilevel energy dialogues will be established for the debate on energy and climate policies. For plans covering the period from 2020 to 2030, Member States should update them by 30 June 2024, at the latest. Status reports will also be produced and should be prepared in order to "ensure transparency towards the Union, other Member States, regional and local authorities, market actors including consumers, any other relevant stakeholders and the general public". In addition, in order to limit the administrative burden for the Member States and the Commission, the Commission will launch an online platform in order to facilitate cooperation with the Member States, promote cooperation between them and facilitate public access to information. Furthermore, the Commission will be assisted by a Climate Change Committee and by an Energy Committee.

6.2 Most Significant Results

6.2.1 *Attainment of the Internal Market for Energy*

The European Council held in Brussels on 23–24 October 2014, which approved the so-called *"Energy-Climate 2030 Package"*, gave entrance to an Energy Union based fundamentally on the conquest of a single and well-connected internal market, as well as the protection of consumers and high interconnection standards. However, to date, the lack of connections is evident in the EU, especially between France and Spain. Although the main Spanish government leaders express their discontent over French nuclear protectionism, their French counterparts emphasize the financing, technical and environmental difficulties. In the words of the expert Gonzalo Escribano, the reproaches are mutual (Escribano, 2014). Achieving an energy union beyond the Pyrenees is a priority. For this reason, the financing of networks through the Pyrenees, including electricity networks, remains a pending issue.

"A regional approach has been and will continue to be decisive for the integration of the European energy market in terms of cross border exchanges as well as security of supply", as indicated in the European Energy Security Strategy (European Commission, 2017a). Thus, the Nordic countries lead in the integration of their markets in the electricity sector (NorPool). The so-called Pentalateral Forum in the Northwest, formed by Belgium, the Netherlands, Luxembourg, Austria, Germany and France works on relevant integration projects in the gas and electricity sectors. Also, the appearance of the PRISMA gas platform, which serves the majority of EU carriers, is also noteworthy.

In relation to the legislation, the early liberalization packages of the Internal Energy Market adopted by the union in 1996–1998, in accordance with the electricity and gas directives (the so-called First Package) and the 2003 Second Package (which, among other things, allowed consumers free choice in their gas and electricity suppliers) were far-reaching. But undoubtedly, the Third Package of the Internal Energy Market "laid the basis for European network planning and investment by creating the requirement for Transmission System Operators (TSOs) to co-operate and elaborate regional and European 10-year network development plans (TYNDP) for electricity and gas in the framework of the European Network of TSOs (ENTSO) and by establishing rules of cooperation for national regulators on cross-border investments in the framework of the Agency for the Cooperation of Energy Regulators (ACER)" (European Commission, 2010).

Other subsequent measures have come hand in hand with other legislative instruments that will strengthen the growth of the internal energy market. These are; Regulation (EU) No. 347/2013 guidelines for trans-European energy infrastructure, Regulation (EU) No. 1227/2011 on wholesale energy market integrity and transparency and the Proposal for a Directive of the European Parliament and of the Council on common rules for the internal market in electricity COM(2016) 864 final/2 (European Commission, 2016a), which allowed for clearer electricity bills, the installation of smart electric meters, a degree of protection for customers in energy poverty (Gouardères, 2019), the arrival of new operators in the market, facilitation of electro-mobility or connection of charging points for electric vehicles. Regulations have also been introduced to clarify the functions of the Distribution Network Managers (DNM) such as the Electricity Regulation COM(2016) 0861 which advocates favoring the cross-border management of electricity networks with the setting up of new regional operational centers, and the Regulation on preparedness against risks COM(2016) 0862. Additionally, the Proposal for a European

Parliament and Council Directive on common rules for the internal electricity market COM(2016) 0864 advocates that electricity has to be guided by undistorted price signals. This initiative aims to prioritize "energy efficiency solutions, and contribute to the goal of becoming a world leader in energy production from renewable energy sources, thus contributing to the Union's target to create jobs, growth and attract investments".

In relation to the ACER (OJEU, 2009a), which was launched in 2011 under the direction of a Director, a Board of Directors, a Board of Regulators and a Board of Appeal, its functions include assisting regulatory authorities at national and regional level, monitoring the internal gas and electricity markets, investigating cases of market abuse and even imposing sanctions, although the latter will remain in the hands of Member States. Two regulations creating cooperation structures of the European Networks of Transmission System Operators (ENTSO) are other additional measures which have been adopted: the first, on the access conditions to the network for cross-border trade in electricity (OJEU, 2009b), and the second, on the access conditions to natural gas transport networks (OJEU, 2009c).

The 2018 edition of the report on energy prices and costs in Europe and its wholesale and retail evolution for electricity, gas and oil products between 2008 and 2017–2018 in Europe, presented by the European Commission COM(2019) 1 final (European Commission, 2019c) is highly relevant because it highlights, among other things, "the important role of international fossil fuel prices in driving energy prices in the EU making the case for pursuing our efforts to decarbonize our energy system. Data also show the impact of dollar-denominated international energy prices on our energy bill, reinforcing the necessity to reduce dependence on fossil fuels and highlighting the benefits of pricing the transactions of energy products in Euros to reduce the uncertainty brought by exchange rate volatility" (European Commission, 2019d). Furthermore, the study analyzes the impact of energy costs on the economy, industry, households of the Union, energy policy, competitiveness (cost of European industry) and the cost of industries with intensive use of energy.

6.2.2 *Improvement of Energy Efficiency*

The essential normative of the Union's policy on energy efficiency is Directive 2012/27/EU, of 25 October 2012, which aims to bolster the main objectives of the Europe 2020 Strategy, by requiring Member States "to set indicative national energy efficiency targets ensuring that the EU reaches its headline target of reducing energy consumption by 20% by 2020" (Gouardères, 2018a). Subsequently, the Commission revised their proposal with the ambitious target of 30% energy efficiency by 2030. Some points highlighted in this regulation are that buildings represent 40% of the final energy consumption of the Union, around 75% of the real estate stock is inefficient from the energy point of view and that it is necessary to increase the pace of building innovation because real estate is the sector with the greatest potential for saving energy. All Member States will carry out a strategy that promotes investments in the renovation of residential and commercial buildings nationally, both public and private, to increase their energy efficiency. Also, according to Article 5 of the aforementioned regulation, the States must exercise an exemplary role and, as of 1 January 2014, they will ensure that "3% of the total floor area of heated and/or cooled buildings owned and occupied by its central government is renovated each year to meet at least the minimum energy performance requirements that it has set in application of Article 4 of Directive 2010/31/EU".

In 2016, the EC proposed a review of the Directive 2012/27/EU through the amendments COM(2016) 76 (OJEU, 2017) and COM(2016) 0765 (OJEU, 2017a), in order to strengthen the energy efficiency of new buildings, reduce the energy consumption of existing ones and extract the maximum potential in terms of energy efficiency in the construction sector. A more recent revision, EU 2018/844, which entered into force as of 9 July 2018, "includes measures that will accelerate the rate of building renovation towards more energy-efficient systems and strengthen the energy performance of new buildings, making them smarter" (European Commission, 2018a). This was the first of the eight legislative acts to be adopted in the Clean Energy for All Europeans Package.

The European Fund for Strategic Investments (EFSI) is one of the financial instruments par excellence which finances projects that promote energy efficiency and renewable energies. In fact, the EC has proposed extending its duration until the end of 2020 and has demanded that "the EIB -European Investment Bank- shall target that at least 40% of EFSI financing under the infrastructure and innovation window supports projects with components that contribute to climate action, in line with the

COP21 -2015 United Nations Climate Change Conference- commitments" (OJEU, 2016a).

6.2.3 *Improve Internal Energy Resources, Especially Renewable Energy*

Renewable sources of energy, namely wind, solar, hydroelectric, oceanic, geothermal, biomass and biofuels have become real alternatives to fossil fuels and therefore directly support the reduction of fossil fuel and GHG emissions, diversifying the supply of energy in the Union and reducing dependence on other markets, especially oil and gas. New and renewable energies form an integral part of the EU's target. In terms of regulation, it is worth mentioning the Directive 2009/28/CE, of 23 April 2009, which includes the need for 20% of the Union's energy consumption to come from renewable energies by 2020 and the 28 Member States must obtain a 10% quota on the fuels used for transport. In addition, all States must draw up a proposal for the achievement of national objectives through the national action plans on renewable energy. In these plans, they will indicate the individual renewable energy targets for the electricity, heating and cooling, and transport sectors, and will indicate whether the EU's sustainability criteria are met. The States will also report their progress every two years, outlining the development of the electrical infrastructure, the electric network, the support systems of biofuels in diesel and gasoline, the supply of biomass, etc.

An additional regulation is that relating to State Aid in the field of environmental protection and energy of 2014–2020, which may be compatible with the internal market (OJEU, 2014) Additionally notable is the strategy of the EC "Energy Roadmap 2050", COM(2011) 885 final, which demonstrates that decarbonization is viable and necessary for environmental, safety and economic issues (OJEU, 2012a). Through the proposed revision of the Renewable Energy Directive, the EU is expected to become a world leader in this sector, with at least 27% of its energy production coming from renewable sources. It also aims to increase renewable energy in the electricity sector, integrate renewable energy in the heating and cooling sector, decarbonize the transport sector and keep customers informed. The Committee of the Regions, an advisory body of the EU, issued an Opinion on renewable energy in 2017, in which it advocated the promotion of cross border cooperation: "Member States shall ensure that support for at least 10% of the newly-supported capacity in each year between 2021 and 2025 and at least 15% of the newly-supported capacity in each year between 2026 and 2030 is open to installations located in other Member States. Cross-border cooperation investments which provide for an appropriate level of interconnections should also be promoted" (OJEU, 2017b). On the other hand, and in the words of Gonzalo Escribano, Director of the Energy Program, Elcano Royal Institute, "Renewable energies are autochthonous energies which produce zero marginal cost: once the investment is made, there is no volatility in prices, no correlation with any fuel, no decline, no emissions. Moreover, they project soft energy power, in the sense of Nye, of the attractive power of an energy model such as the European one, based on sustainability and committed to the fight against climate change" (Escribano, 2015). Concerning other specific resources such as biomass and biofuels, maritime wind energy and ocean energy, other objectives to be reached in 2020 are as follows: 10% of transport fuels must come from renewable energies, and suppliers will be obliged to include the intensity values of production from the fuels supplied by the Member States so that any increase in the volume of production of high-carbon fuels must be accompanied by efforts to reduce emissions in other sectors and to ensure a 6% reduction in GHG emissions from transport by 2020.

An exhaustive analysis of the current state of affairs and progress of the different EU countries regarding issues related to the energy market, or regarding the circular economy, renewable energies and resource efficiency, reveals some interesting and certainly disparate data. For example, "Sweden is already outperforming its 2020 target of 49%". The country intends to continue to mobilize investments to achieve the 2030 EU RES target of 32%. Austria, ambitious in its climate policy, is on track "to meet the renewable target for 2020 but risks missing the energy efficiency and greenhouse gas emission reduction targets" (European Commission, 2019e). In the case of the Iberian Peninsula, "with the completion of the Val de Saône gas pipeline in France in October 2018, Spain has improved its access to the internal market. Additional gas interconnections ('MIDCAT' and 'STEP' as well as a third interconnection point between Spain and Portugal) have received the status of Projects of Common Interest and are being assessed". There has also been an improvement in multi-modal connections on the rail. "However, some Spanish transport infrastructures (ports, high-speed rail connections and airports) operate well below their capacity, and the share of freight transported by rail remains low at 5.1% in 2016. Besides, "the Trans-European Transport Network (TEN-T) Atlantic and Mediterranean corridors remain incomplete" (European Commission, 2019f). Bulgaria stands out among the countries that require huge investments in the field of energy and climate. This country "is expected to adopt the final version of its Integrated National Energy and Climate Plan by 31st December 2019, as required by the Energy Union Governance Regulation, with an overview of investment needs for the different dimensions of the Energy Union until 2030" (European Commission, 2019g).

6.2.4 *Increase the Security of Energy Supply*

The EU imported 53.6% of the energy it consumed in 2016. Forecasts are that this trend will continue to rise in the coming years. Although the supply of fossil fuels such as coal and petroleum does not pose great inconveniences (In the Directive 2009/119 CE of the Council, of 14th September 2009 (OJEU, 2009d). States are obliged to maintain a minimum level of reserves of crude oil or petroleum products – minimum requirement of 90 days of average daily net imports or 61 days of daily average domestic consumption or, in other words, the emergency supply according to the International Energy Agency -AIE-), it is in the field of natural gas where the real problems lie.

The main gas suppliers of the EU are Russia providing 39%, Norway 33% and North Africa (Algeria and Libya) 22%. Other sources represent only 4%. However, the Russian Federation is not only the main supplier of gas but also a major supplier of oil, coal and nuclear fuel (European Political Strategy Centre, 2018). For this reason, the situation becomes more convoluted for the EU when faced with Russian assertiveness in international politics, its actions and behavior toward Ukraine and the illegal annexation of Crimea in 2014 which violated international law. This situation has led the 28 states of the Union to establish sanctions against Russia and has been a headache for the EU with respect to Ukraine, a country with a strategic location and the main route for transporting gas to Europe. After crossing Ukraine and Belarus, Poland, the Czech Republic and Slovakia are the main transit countries for Russian gas, to which Germany will be added when Nord Stream 2 is completed, a very controversial project which is not viewed favorably by Sweden, the Baltic States or Poland. Although Germany sells it as an apparently simple commercial agreement, it is clear that this option will result in the concentration of 80% of the supply of Russian gas in a single route and that it would be controlled mainly by Gazprom (Dempsey, 2017).

Russia is an unreliable partner who uses the energy supply as a political tool. Therefore, for the EU the diversification of the energy supply is a must. For this purpose, "the Southern Gas Corridor aims to expand infrastructure that can bring gas to the EU from the Caspian Basin, Central Asia, the Middle East, and the Eastern Mediterranean Basin. Initially, approximately 10 billion cubic meters (bcm) of gas will flow along this route when it opens in 2019-2020. Given the potential supplies from the Caspian Region, the Middle East and the East Mediterranean, however, the EU aims to increase this amount up to 80 to 100 bcm per year in the future" (European Commission, 2019h). Additionally, the EU wants to create a Mediterranean gas hub in the South of Europe in line with this diversification of suppliers and routes. The EU is therefore engaged in an active energy dialogue at a political level with North African and Eastern Mediterranean partners. The acquisition of liquefied natural gas (LNG) from the US, Australia or Qatar would be another line of diversification, which would provide the EU with a new geopolitical scenario not so exposed to the risk of a preponderant Russian energy supply. The EU also recognizes the need to revitalize energy diplomacy and international dialogue with key partners such as Turkey, Algeria, Turkmenistan, Azerbaijan Africa, the Middle East and other possible future suppliers.

The Gas Coordination Group, involving the Member States, regulators and all interested parties, has proven to be an effective pan-European platform for the exchange of information between experts in energy security and the coordination of actions between Member States. Of special relevance are the works that promote greater cross-border cooperation, for example, through the development of risk assessments (stress tests) and security plans for supply at regional and EU levels, through the development of a regulatory framework for gas storage that recognizes its strategic importance in security of supply or by a more precise definition of "protected consumers". Likewise, the maintenance of strategic energy infrastructures and the protection of critical infrastructures such as gas and electricity transportation systems are critical issues. In fact, the 2014 European Energy Security Strategy (European Commission, 2014), which is part of the climate and energy policy framework for 2030 and which is consistent with European competitiveness and industrial policy objectives, points out that decisions to support critical infrastructure projects (such as Nord Stream, South Stream, The Trans-Adriatic Gas Pipeline (TAP) or a Baltic LNG terminal) should be discussed at European or regional level to ensure that the decisions of one Member State do not adversely affect the security of supply of another Member State.

The most relevant legal acts that the Community institutions have put in place in order to guarantee energy security are:

- *Regulation (EU) no. 994/2010*, whose purpose is to guarantee the security of gas supply and advocates strengthening mechanisms for prevention and response to crises. The Commission has proposed extending the scope of application of Directive 2009/73 / EC – the Gas Directive – to gas pipelines to and from third countries, both existing and in the future (European Commission, 2017b).
- *Regulation 2017/1938* (OJEU, 2017c), developed in response to the crisis in Ukraine, provides for greater regional cooperation, preventive action plans and emergency plans, as well as a solidarity mechanism to ensure the security of the gas supply.
- *Decision (EU) 2017/684* of the European Parliament and the Council of 5 April 2017 establishes an information exchange mechanism with regard to intergovernmental agreements and non-binding instruments between the Member States and third countries in the sector of energy, and repealed Decision No. 994/2012/EU (OJEU, 2017d), which had been a milestone in guaranteeing the conformity of the intergovernmental agreements on energy with the common energy policy. As noted in the 2017 Decision:
  - "...Decision nº 994/2012/EU proved ineffective in terms of ensuring compliance of intergovernmental agreements with Union law. That Decision mainly relied on the assessment of intergovernmental agreements by the Commission after they were concluded by the Member States with a third country. Experience gained in the implementation of the Decision 994/2012/EU demonstrated that such an ex-post assessment does not fully exploit the potential for ensuring

compliance of intergovernmental agreement with Union law. In particular, intergovernmental agreements often contain no appropriate termination or adaptation clauses which would allow Member States to bring the intergovernmental agreement in compliance with Union law within a reasonable period of time. Furthermore, the positions of the signatories have already been fixed, which creates political pressure not to change any aspect of the agreement".

- Member States will be required to notify the Commission in advance of draft intergovernmental agreements on oil or gas before they are legally binding for the parties. Member States shall refrain from entering into intergovernmental agreements relating to gas or oil, or an intergovernmental agreement relating to electricity where a Member State has requested the prior evaluation of the Commission and shall wait until the Commission has informed them of its assessment. Only intergovernmental agreements relating to the purchase, trade, sale, transit, storage or supply of energy in at least one Member State, or the construction or exploitation of energy infrastructure physically connected to at least one Member State, must be notified. If there is any doubt, the Member States will consult the Commission. However, this decision should not create obligations with respect to agreements between companies. In any case, Member States should be able to communicate voluntarily to the Commission the trade agreements that are explicitly mentioned in intergovernmental agreements or non-bidding instruments.

- *Communication from the Commission on an EU strategy for liquefied natural gas (LNG) and gas storage COM(2016) 049* (European Commission, 2016b). The EU wanted to examine the potential of LNG and its storage capacities, in order to diversify the EU gas system and contribute to a more resilient, competitive and secure supply in accordance with the ambition of the Communication of the European Commission which is titled *"Framework Strategy for a Resilient Energy Union with a Forward-Looking Climate Change Policy COM(2015) 080"* (European Commission, 2015). Regional cooperation is being developed through the fundamental role played by the so-called High-Level Groups (European Commission, 2017c), which "have contributed in particular to the prioritization of key projects of common interest in the region". The scope of certain High-Level Groups has been extended "to cover wider aspects of energy policy, notably energy markets, renewables generation and energy efficiency". As it appears in COM(2016) 049, these groups have agreed on a limited number of key projects of common interest (PCIs) that should be implemented as a matter of priority and urgency:

- The High-Level Group on Gas Connectivity in Central and South-Eastern Europe (CESEC Group) aims to improve the access of these countries through the two main corridors: "from the Krk terminal towards the east and from Greece to the north".
- "The Baltic Energy Market Interconnection Plan (BEMIP) group identified six key priority projects that contribute to LNG and storage access in the region by connecting the three Baltic States and Finland to the European network".
- The High-Level Group on Interconnections for Southwestern Europe formed by Spain, France and Portugal is also paying off. It identified "two projects that would serve to eliminate bottlenecks and connect regional markets", mentioned in the Madrid Declaration after the Energy Interconnections Links Summit Spain–France–Portugal–EC–EIB, Madrid, 4 March 2015 (Rajoy et al., 2015), and included in the Regulation (EU) 2016/89 of 18 November 2015 (OJEU, 2016b): project 5.4 – the "3rd interconnection point between Portugal and Spain", and project 5.5 – the completion of the "Eastern Axis Spain-France - interconnection point between Iberian Peninsula and France at Le Perthus, including the compressor stations at Montpellier and St. Martin de Crau (currently known as 'MidCat')". A third project has been recently included by the EC in the list of PCI in the Regulation (EU) 2018/540 of 23 November 2017: the South Transit East Pyrenees (currently known as "STEP") (OJEU, 2018b).

6.2.5 *Research & Development & Innovation (R&D&I) Projects*

The main financial instrument of the EU to promote research and innovation in energy issues is the Horizon 2020 Program, which covers the period 2014–2020. Approved by the European Commission in 2013, it is structured in three fundamental pillars: excellent science, industrial leadership and social challenges. More than 5,931 million Euros have been designated for the achievement of clean, safe, efficient, competitive and low-carbon energy. Investments in R&D&I should include the entire technological supply chain, from materials to manufacturing.

The *European Strategic Energy Technology (SET) Plan* COM(2006) 0847 final (European Commission, 2006), adopted by the Commission on 22 November 2007, aims to transform the energy system gradually in the coming decades, trying to "embrace all aspects of technological innovation, as well as the policy framework required to encourage business and the financial community to deliver and support the efficient and low carbon technologies that will shape our common future". The document COM(2015) 6317 gathers the Opinion of the European Economic and Social Committee (EESC) on the "Communication from the Commission - Towards an integrated Strategic Energy Technology (SET) Plan: accelerating the European energy system transformation" (OJEU, 2016c). In the EESC's view, "the most important task is the technical and scientific development of technologies and innovation, and the promotion of factors that encourage new ideas and concepts". This advisory body of the Union is in favor of an increase in the financial budget and of giving more participation to the interested parties throughout the entire chain of research and innovation. Among its more concrete observations, it welcomed the use of hydrogen for the storage of electricity (although, it never hurts to remember that hydrogen is as renewable as the energy sources used for its generation!) and it stressed the important role played by cities in the decarbonization of the economy of the Union. Another issue pointed out by this body is the recycling of lithium-ion batteries and the recycling of batteries of electric and hybrid vehicles, and it also called for the combination of sustainable technologies such as "advanced biofuels, hydrogen and alternative liquids and gaseous fuels, including LNG". Regarding nuclear energy, there is no unanimity in the EU and the policy is heterogeneous. With respect to nuclear energy, and according to the EESC, "the EU's policy is anything but unified". However, "...Advanced new reactors under construction can lead to a nuclear renaissance, so the nuclear revival seems to be a fact despite brief hesitation. Time will tell us whether or not the EU can cut back on the share of nuclear power in its energy mix, but so far the wheels are turning".

The EU strategy that promotes innovation in energy is defined in the Communication of the Commission entitled "Energy Technologies and

Innovation" COM(2013) 253, published on 2 May 2013. In this document, the Commission requests that the European Parliament and the council "reaffirm their support to the SET Plan as part of Europe's Energy and Climate Change policies and its reinforcement to energy technology and innovation development as set out in this Communication" (OJEU, 2013).

6.2.6 *Main Funding Instruments*

Energy has received significant budgetary injections precedent from different funding instruments in the current Multi annual Financial Framework (MFF) 2014–2020. It has received around 2,000 million Euros from the European Regional Development Fund *(ERDF)*, providing support to intelligent energy storage and transport systems, favoring electricity and gas infrastructures, and supporting the "low-carbon economy", one of the four priority areas of the current MFF. The Cohesion Funds have also financed environmentally friendly energy projects. In fact, more than 350,000 million Euros from regional and cohesion funds have been designated for energy projects. Research in the energy field has also received relevant funding. The International Experimental Thermonuclear Reactor (ITER) has a budget of 3,000 million Euros. Neither EURATOM Nuclear Research Program nor the 2018–2020 Work Program "Secure, clean and efficient energy", should be disregarded, with more than 2,300 million Euros.

The *Connecting Europe Facility (CEF)* was created under the MFF 2014–2020 and provides financial support to PCIs in the areas of trans-European transport, telecommunications and energy networks. The *PCIs* (European Commission, 2019i), key cross-border infrastructure projects in line with the objectives of climate policy include electrical and gas connections, electricity and gas storage and liquefied gas re-gasification stations, smart grids, oil etc. They look for a progressive approach to the Union of the Energy and can opt to the financial aid of the Mechanism "Connect Europe" (MCE). The Commission updates the status of their situation through a platform (European Commission, 2019j). In 2013, the first list with key projects was prepared. The latest list was submitted in November 2017 (OJEU, 2018c). The list is updated every two years and corresponds to the following projects:

1. Priority Corridor Northern Seas Offshore Grid (NSOG).
2. Priority Corridor North-South Electricity Interconnections in Western Europe ("NSI West Electricity").
3. Priority Corridor North-South Electricity Interconnections in Central Eastern and South Europe ("NSI East Electricity").
4. Priority Corridor Baltic Energy Market Interconnection Plan ("BEMIP Electricity").
5. Priority Corridor North-South Gas Interconnections in Western Europe ("NSI West Gas").
6. Priority Corridor North-South Gas Interconnections in Central Eastern and South-Eastern Europe ("NSI East Gas").
7. Priority Corridor Southern Gas Corridor ("SGC").
8. Priority Corridor Baltic Energy Market Interconnection Plan in Gas ("BEMIP Gas").
9. Priority Corridor Oil Supply Connections in Central Eastern Europe ("OSC").
10. Priority Thematic Area Smart Grids Deployment.
11. Priority Thematic Area Electricity Highways (NSOG; NSI West Electricity; NSI East Electricity and BEMIP Electricity.
12. Cross-border carbon dioxide network.

In the framework of the CEF for 2014–2020, a budget of 5,350 million Euros has been allocated to trans-European energy infrastructures in order to help integrate European energy markets and diversify their sources. In the words of the Commissioner for Action for Climate and Energy, Arias Cañete, "Modern and reliable energy infrastructure is essential to allow energy to flow freely across Europe. These projects will help us integrate our energy markets, diversify energy sources and routes, and end the energy isolation of some Member States. They will also boost the level of renewables on the grid, slashing carbon emissions. Europe's energy transformation will require investments worth billions in strategic infrastructure" (European Commission, 2015a). It is also relevant to highlight the EFSI, an initiative of the EC and the EIB. This fund, as mentioned above, aims to attract investment in key sectors such as renewable energy and resource efficiency.

Conclusion

To talk about the "Union of Energy" is to talk not only about energy but also about the climate, the European economy and investments. According to data provided by the EC in the "Fourth report on the State of the Energy Union", "the financial sector in the EU has the potential to deliver the annual investment needs of almost 180,000 million Euros to achieve the EU's climate and energy targets by 2030". Only through public–private collaboration and its attendant strategies can one go toward a clear path in the trinomial: energy, innovation and sustainability.

The transformations that are required by society in general, companies and other main actors involved in order to achieve the 2020, 2030 and 2050 objectives and go toward "an economically neutral economy" are enormous and the processes of "adaptation" do not look easy and are full of uncertainty. Even so, the progress made in recent years is not insignificant. The Juncker Commission, in this transition that we are living – from decarbonization to cleaner energy – has been decisive and has advocated modernizing the European economy and industry, linking the Union to the Paris Agreement.

The crux of the energy policy is made up of a huge number of measures and European regulations which, in combination with no less numerous amounts of national legislation, are aimed at strengthening the internal energy market, guaranteeing the security of energy supply, promoting energy efficiency along with clean and renewable energies, reducing GHG emissions, promoting research and innovation, promoting regional cooperation and collaboration with third states.

The guarantee of supply and its diversification is one of the workhorses of the Union. Today, the main energy suppliers for the EU are Norway, Iran, Algeria, Turkey, Egypt and Russia, this latter source always surrounded by controversy over the political use of its resources, as in the case of the crisis with Ukraine. The Southern Gas Corridor will start to be effective in 2020 and there are many expectations for Spain and Portugal to become a hub for Europe, capable of diversifying the supply of gas to Western Europe. Currently, according to the data of the EC "all Member States, except one, have access to two independent sources of gas and, if all projects in progress are executed according to the calendar, all countries except Malta and Cyprus will have access to three gas sources by 2022 and 23 Member States will have access to the global liquefied natural gas market".

Finally, although emissions have decreased in practically all sectors in recent years (except in transport), the rate of growth of renewable energies has slowed since 2014, affecting Belgium, France, Ireland, Luxembourg, France The Netherlands, Poland and Slovenia. In a time of political uncertainties, the commitment of countries to reducing the impact on the environment can be affected by the appearance of weak governments that try to offer citizens simple solutions to complex problems. We will see its effects when the Commission publishes its next report on the state of the Energy Union before October 2020.

References

Dempsey, J. (2017) *Judy Asks: Is Europe Too Dependent on Russian Energy?* Carnegy Europe Web page, available at: https://carnegieeurope.eu/strategiceurope/71507

Escribano, G. (2014) Une Union de l'énergie au-delà des Pyrénées, *Le Monde 25 October 2014.*

Escribano, G. (2015) ¿Qué nos deparará 2015 en energía?, ARI 1/2015 - 5/1/2015, *Elcano Royal Institute web page.*

European Commission (2006) COM/2006/0847 final, *Communication from the Commission to the Council, the European Parliament, the European Economic and Social Committee and the Committee of the Regions - Towards a European strategic energy technology plan.*

European Commission (2010) COM/2010/677 final, *Commission Staff Working Document Impact Assessment Accompanying document to the Communication from the Commission to the European Parliament, the Council, the European Economic and Social Committee and the Committee of the Regions - Energy infrastructure priorities for 2020 and beyond - A Blueprint for an integrated European energy network.*

European Commission (2014) COM/2014/0330 final, *Communication from the Commission to the European Parliament and the Council - European Energy Security Strategy.*

European Commission (2015) COM/2015/080 final, *Communication from the Commission to the European Parliament, the Council, the European Economic and Social Committee, the Committee of the Regions and the European Investment Bank - A Framework Strategy for a Resilient Energy Union with a Forward-Looking Climate Change Policy.*

European Commission (2015a) *Press Release - Commission unveils key energy infrastructure projects to integrate Europe's energy markets and diversify sources, Brussels, 18 November 2015.*

European Commission (2016a) COM/2016/0864 final/2, *Proposal for a Directive of the European Parliament and of the Council on common rules for the internal market in electricity (recast).*

European Commission (2016b) COM/2016/049 final, *Communication from the Commission to the European Parliament, the Council, the European Economic and Social Committee and the Committee of the Regions on an EU strategy for liquefied natural gas and gas storage.*

European Commission (2016) *Press Release - Clean Energy for All Europeans – unlocking Europe's growth potential, Brussels, 30 November 2016.*

European Commission (2017) *The EU and Energy Union and Climate Action*. Publications Office of the European Union: Luxembourg.

European Commission (2017a) COM/2018/330 final, *Communication from the Commission to the European Parliament, the Council, the European Economic and Social Committee and the Committee of the Regions - A Europe that protects: Clean air for all.*

European Commission (2017b) COM/2017/0660 final, *Proposal for a Directive of the European Parliament and of the Council amending Directive 2009/73/EC concerning common rules for the internal market in natural gas.*

European Commission (2017c) COM/2017/718 final, *Communication from the Commission to the European Parliament, the Council, the European Economic and Social Committee and the Committee of the Regions - Communication on strengthening Europe's energy networks.*

European Commission (2018) *Standard Eurobarometer 90 - Autumn 2018.*

European Commission (2018a)*News - New Energy Performance in Buildings Directive comes into force on 9 July 2018.*

European Commission (2019) *Fact Sheet - The Energy Union Five Years On - The Juncker Commission Delivers On Its Energy Union Priority.*

European Commission (2019a) *EU Emissions Trading System (EU ETS)*, available at: https://ec.europa.eu/clima/policies/ets_en, consulted in April 2019.

European Commission (2019b) *NER 300 Programme,* available at: https://ec.europa.eu/clima/policies/innovation-fund/ner300_en

European Commission (2019c) COM/2019/1 final, *Report from the Commission to the European Parliament, the Council, the European Economic and Social Committee and the Committee of the Regions – Energy prices and costs in Europe.*

European Commission (2019d) SWD/2019/1 final, *Commission Staff working document accompanying the document Report from the Commission to the European Parliament, the Council, the European Economic and Social Committee and the Committee of the Regions – Energy prices and costs in Europe.*

European Commission (2019e) SWD/2019/1019 final, *Commission Staff Working Document - Country Report Austria 2019 - Accompanying the document Communication from the Commission to the European Parliament, the European Council, the Council, the European Central Bank and the Eurogroup 2019 European Semester: Assessment of progress on structural reforms, prevention and correction of macroeconomic imbalances, and results of in-depth reviews under Regulation (EU) No 1176/2011.*

European Commission (2019f) SWD/2019/1008 final, *Commission Staff Working Document - Country Report Spain 2019 - Accompanying the document Communication from the Commission to the European Parliament, the European Council, the Council, the European Central Bank and the Eurogroup 2019 European Semester: Assessment of progress on structural reforms, prevention and correction of macroeconomic imbalances, and results of in-depth reviews under Regulation (EU) No 1176/2011.*

European Commission (2019g) SWD/2019/1001 final, *Commission Staff Working Document - Country Report Bulgaria 2019 - Accompanying the document Communication from the Commission to the European Parliament, the European Council, the Council, the European Central Bank and the Eurogroup 2019 European Semester: Assessment of progress on structural reforms, prevention and correction of macroeconomic imbalances, and results of in-depth reviews under Regulation (EU) No 1176/2011.*

European Commission (2019h) *Diversification of gas supply sources and routes*, Web page of the European Commission, available at: https://ec.europa.eu/energy/en/topics/energy-security/diversification-of-gas-supply-sources-and-routes, consulted in April 2019.

European Commission (2019i) *Projects of Common Interest*, Web page of the European Commission, available at: https://ec.europa.eu/energy/en/topics/infra-structure/projects-common-interest, consulted in April 2019.

European Commission (2019j) *Projects of common interest – Interactive map*, Web page of the European Commission, available at: http://ec.europa.eu/energy/infrastructure/transparency_platform/map-viewer/main.html, consulted in April 2019.

European Parliamentary Research Service (2018) *Briefing - EU Policies - Delivering for Citizens - Energy Supply and Security.* European Parliament Think Tank.

European Political Strategy Centre (2018) *10 Trends Reshaping Climate and Energy.* European Commission.

General Secretariat of the European Council (2014) *European Council, 23/24 October 2014 - Conclusions.*

Gouardères, F. (2018) *Fact Sheets on the European Union - Energy Policy: General Principles.* European Parliament.

Gouardères, F. (2018a) *Fact Sheets on the European Union – Energy Efficiency.* European Parliament.

Gouardères, F. (2019) *Fact Sheets on the European Union – Internal Energy Market.* European Parliament.

OJEU (2009) L140, *Decision No 406/2009/EC of the European Parliament and of the Council of 23 April 2009 on the effort of Member States to reduce their greenhouse gas emissions to meet the Community's greenhouse gas emission reduction commitments up to 2020.*

OJEU (2009a) L211, *Regulation (EC) No 713/2009 of the European Parliament and of the Council of 13 July 2009 Establishing an Agency for the Cooperation of Energy Regulators.*

OJEU (2009b) L211, *Regulation (EC) No 714/2009 of the European Parliament and of the Council of 13 July 2009 on conditions for access to the network for cross-border exchanges in electricity and repealing Regulation (EC) No 1228/2003.*

OJEU (2009c) L211, *Regulation (EC) No 715/2009 of the European Parliament and of the Council of 13 July 2009 on conditions for access to the natural gas transmission networks and repealing Regulation (EC) No 1775/2005.*

OJEU (2009d) L265, *Council Directive 2009/119/EC of 14 September 2009 imposing an obligation on Member States to maintain minimum stocks of crude oil and/or petroleum products.*

OJEU (2012) C326, *Consolidated version of the Treaty on the Functioning of the European Union.*

OJEU (2012a) C229, COM/2011/885 final, *Opinion of the European Economic and Social Committee on the 'Communication from the Commission to the European Parliament, the Council, the European Economic and Social Committee and the Committee of the Regions - Energy Roadmap 2050'.*

OJEU (2013) C67, COM(2013) 253 final, Opinion of the European Economic and Social Committee on the 'Communication from the Commission to the European Parliament, the Council, the European Economic and Social Committee and the Committee of the Regions on Energy technologies and innovation'.

OJEU (2014) C200, *Communication from the Commission - Guidelines on State aid for environmental protection and energy 2014–2020.*

OJEU (2016) C202, *Consolidated version of the Treaty on European Union - Protocol (No 37) on the Financial consequences of the expiry of the ECSC Treaty and on the research fund for coal and steel.*

OJEU (2016a) C246, COM/2016/0860 final, *Opinion of the European Economic and Social Committee on the 'Communication from the Commission to the European Parliament, the Council, the European Economic and Social Committee, the Committee of the Regions and the European Investment Bank - Clean Energy For All Europeans'.*

OJEU (2016b) C75, COM/2016/597 final, Opinion of the European Economic and Social Committee on the proposal for a Regulation of the European Parliament and of the Council amending Regulations (EU) No 1316/2013 and (EU) 2015/1017 as regards the extension of the duration of the European Fund for

Strategic Investments as well as the introduction of technical enhancements for that Fund and the European Investment Advisory Hub.

OJEU (2016c) L19, *Commission Delegated Regulation (EU) 2016/89 of 18 November 2015 amending Regulation (EU) No 347/2013 of the European Parliament and of the Council as regards the Union list of projects of common interest.*

OJEU (2017) C246, COM/2016/761 final, *Opinion of the European Economic and Social Committee on the 'Proposal for a directive of the European Parliament and of the Council amending Directive 2012/27/EU on energy efficiency'.*

OJEU (2017a) C246, COM/2016/765 final, *Opinion of the European Economic and Social Committee on the proposal for a directive of the European Parliament and of the Council amending Directive 2010/31/EU on the energy performance of buildings.*

OJEU (2017b) C342, *Opinion of the European Committee of the Regions - Renewable energy and the internal market in electricity.*

OJEU (2017c) L280, *Regulation (EU) 2017/1938 of the European Parliament and of the Council of 25 October 2017 concerning measures to safeguard the security of gas supply and repealing Regulation (EU) No 994/2010.*

OJEU (2017d) L99, *Decision (EU) 2017/684 of the European Parliament and of the Council of 5 April 2017 on establishing an information exchange mechanism with regard to intergovernmental agreements and non-binding instruments between Member States and third countries in the field of energy, and repealing Decision No 994/2012/EU.*

OJEU (2018) L156, PE/3/2018/REV/2, *Regulation (EU) 2018/842 of the European Parliament and of the Council of 30 May 2018 on binding annual greenhouse gas emission reductions by Member States from 2021 to 2030 contributing to climate action to meet commitments under the Paris Agreement and amending Regulation (EU) No 525/2013.*

OJEU (2018a) L328, PE/55/2018/REV/1, *Regulation (EU) 2018/1999 of the European Parliament and of the Council of 11 December 2018 on the Governance of the Energy Union and Climate Action, amending Regulations (EC) No 663/2009 and (EC) No 715/2009 of the European Parliament and of the Council, Directives 94/22/EC, 98/70/EC, 2009/31/EC, 2009/73/EC, 2010/31/EU, 2012/27/EU and 2013/30/EU of the European Parliament and of the Council, Council Directives 2009/119/EC and (EU) 2015/652 and repealing Regulation (EU) No 525/2013 of the European Parliament and of the Council.*

OJEU (2018b) L90, *Commission Delegated Regulation (EU) 2018/540 of 23 November 2017 amending Regulation (EU) No 347/2013 of the European Parliament and of the Council as regards the Union list of projects of common interest.*

OJEU (2018c) L90, *Commission Delegated Regulation (EU) 2018/540 of 23 November 2017 amending Regulation (EU) No 347/2013 of the European Parliament and of the Council as regards the Union list of projects of common interest.*

Presidency of the European Union (2007) *Presidency Conclusions of the Brussels European Council.*

Rajoy, M., Hollande, F., Passos, P. and Juncker, J.C. (2015) *Madrid Declaration, Energy Interconnections Links Summit, Spain-France-Portugal-European Commission-EIB*. La Moncloa, Gobierno de España: Madrid.

CHAPTER 7

Implementing Climate Security in the European Union

Rosa Giles-Carnero

We know what our principles, our interest and our priorities are. This is no time for uncertainty: our Union needs a Strategy. We need a shared vision, and common action.

–Federica Mogherini in EEAS (2016)

Introduction

Successive reports by the Intergovernmental Panel on Climate Change (IPCC) have described the many transversal effects of changes taking place in the global climate system, such as risks to the stability of social and economic systems. The reports have assessed the potential impact of climate change on different regions, and on the world in general, and warn that the vulnerability of territories and societies to the effects of climate change could lead to destabilization and conflict. The IPCC has analyzed the effects of climate change from the security perspective, by applying a human security concept to reflect on the complex interrelation between the environmental effects of climate change and the maintenance of peace and security. In particular, this approach can be observed in the latest report approved by the IPCC in 2014, which included a section dedicated to the analysis and assessment of the connection between climate change and human security (Adger et al., 2014). In assessing the vulnerability of territories and societies, and analyzing the risks that the effects of climate change would generate, the IPCC concludes by calling for action in order to adapt to a new set of climate circumstances.

These studies by, and warnings from, the IPCC add to growing concern about the impact of climate change on security at national and international level, and the question of climate security now forms part of security agendas. Various states, mainly in the West, have incorporated climate change into their security strategies, in line with actions developed in this area by the United Nations collective security system. The United Nations Security Council has intensified its analysis of the question of climate change with two open debates (Security Council, 2007; United Nations Security Council, 2011) and two Arria Formula meetings, which could be particularly relevant in the eventual inclusion of some of the effects of climate change in declarations on threats to international peace and security, in accordance with Article 39 of the Charter of the United Nations. This development has been termed by the doctrine as a securitization process, by which an environmental question such as climate change is addressed from a perspective that emphasizes its connection to security. In analyses of international relations, the work on securitization theory by the Copenhagen School of Security Studies is of particular importance (Buzan et al.,1998) and has influenced publications that refer specifically to climate change (Brauch et al. 2012; Page 2000; and Scott 2012); although it needs to be stated that after a wide-ranging debate between critics and counter-critics, a consensual and comprehensive theory on securitization is far from being achieved (Balzacq, 2010). This process highlights the many effects that climate change can have on stability in territories and societies and is observed on the different levels of decision-making related to actions on climate change.

Reflection on climate change management from the security perspective offers a new approach to the analysis of governance when confronting the effects of this environmental phenomenon. This focus means that the environmental risk arising from modifications to the climate is conceptually transformed into a threat to territories and socio-economic systems, and thus, to the maintenance of peace and security. Although an approach based on the securitization of climate change has its critics, it also provides a new opportunity to incentivize the adoption of actions to mitigate, and adapt to, climate change within the array of national systems and at international level. The development of the international regime on climate change is a good example, showing how concern on questions of security can encourage states to adopt specific obligations on climate change, to the extent that it is seen as an incentive to promote climate diplomacy to empower an international system capable of halting an extremely dangerous rise in global temperatures.

The member states of the European Union (EU) have not remained aloof from this securitization process, with climate change figuring ever more prominently in their national agendas and security strategies. For example, the United Kingdom has been particularly prominent in this field, as pioneer in initiating a securitization process in climate change management over a decade ago (Harris, 2012). Given such activity among its member states, the EU could not remain unmoved by the trend to address climate change from the security planning perspective, especially when presented with specific data on the vulnerability of the European territory by the European Environment Agency. In the most recent report dating from 2017, the Agency provides a broad analysis of the environmental and social impact on Europe and recommends effective actions for tackling this problem (European Environment Agency, 2017). As a result, the EU has assessed the impact of the effects of climate change on the security of its territories and populations, and thus developed an approach to a concept of climate security that assumes the need to consider the many repercussions of climate change on Europe and worldwide. As successive EU security strategies have developed, they have incorporated the evaluation of the connection between climate change management and peacekeeping and security, and have designed various lines of action in this field.

The notion of climate security thus provides a new perspective for development in the EU in terms of analysis and planning of actions to counter climate change. That said, in the European setting, this evolution, which has seen the incorporation of concern regarding the connection between climate change and security, must be framed within more general actions to face down this environmental phenomenon, and which cover both internal and external aspects. Internally, the EU has developed broad complex actions to mitigate the effects of climate change, setting objectives for 2020 to reduce greenhouse gas emissions by 20% and to increase by the same percentage the use of renewable energies and energy efficiency. To achieve these objectives, the EU has rolled out extensive legal initiatives based on the creation of a European market covering rights to emit greenhouse gases, later gathered together as the climate and energy package. Although the current package was designed for application up to 2020, more ambitious objectives were adopted for 2030, and are been planning for 2050. The 2030 climate and energy framework includes new targets of at least 40% cuts in greenhouse gas emissions, 32% share for renewable energy, and 32.5% improvement in energy efficiency, to which is added that a new long-term strategy for 2050 should allow the EU to adopt and submit more ambitious legislative actions. This has been matched in recent years by action in the field on adaptation, which has seen comparatively less legislative activity, but which the European Commission adopted in 2013 as the EU Strategy on adaptation to climate change (European Commission Communication (16 April 2013). This text was presented as a general framework for planning the development of regulations on adaptation to enable the EU territory and societies to increase their resilience to climate change.

On an international level, the EU has aimed to provide leadership by the creation of a confidence framework based on defending a European model that assumes important commitments to the climate while maintaining levels of growth and quality of life of the European citizen. This fundamentally connects the EU's internal action to its foreign policy as the European negotiating strategy is based on the idea of sustaining its diplomatic position by example, in having taken on significant commitments to mitigate the effects of climate change. From this perspective, European climate policy is no mere compliance with an international obligation but is a relevant contribution to the advancement of an international legal regime and the source of the credibility of the European position in negotiations at the Conference of the Parties. This strategy has not prevented the EU's position of leadership on climate change from suffering important setbacks, yet the EU's External Common Security Policy continues to view playing a significant role in international climate negotiations as a fundamental objective.

It can be clearly seen, therefore, that the process by which the EU added climate change to its security agenda is framed within a general action in which the mitigation of, and adaptation to, climate change has become the fundamental core of its strategy both in internal development and in its international negotiating position. Europe 2020 strategy, launched in 2010, as a general EU development strategy clearly shows this idea (European Commission, 2010). In Europe's 2020 strategy, the action on climate change is positioned as one of the fundamental lines of planning for growth and progress in Europe. With this general framework in mind, the recognition of the relevance of climate security not only emphasizes the transversal nature of the effects of climate change, highlighting the most potentially catastrophic effects for the territories and socio-economic systems, but also establishes a parameter for analysis and action in which new elements are presented for the development of internal legal action on climate change, and a negotiating position in international climate forums. From this perspective, the aim of climate security is just one more aspect to show how EU activity on climate change is significantly shaping its internal policies and the deployment of its external action.

Taking this reflection as a framework, the following sections will analyze how the EU pursues the objective of climate security and its implications for the development of its internal and international activities in the face of climate change. This analysis will consist of two parts. The first will analyze the evolution of the climate security concept within the framework of the EU Common Foreign and Security Policy. The 2003 Security Strategy, which initially made no reference to climate change, will be analyzed, along with the 2016 Global Strategy, in which climate change figures prominently. This first section will describe the progressive conceptualization of the notion of climate security and its connection to international practice and its development by EU member states.

The second section will reflect on the actions to be developed in order to achieve the objective of climate security. The challenge for the concept of climate security included in the strategies is that it is to provide for specific relevant actions. The complexity of this field is considerable because it assumes there is good interconnection between the methodology of security and the methodology of the environment developed through actions to mitigate and adapt to, climate change. It is difficult to translate this notion into concrete measures that will have a significant environmental impact and a specific influence on European law and action in international climate negotiations. Definitively, the problems in implementing climate security are largely due to deficiencies in the process of European integration, hence final conclusions will include a reflection on the impact of the objective of implementing climate security on the general development of the EU committed to action on climate change both inside and outside the European system.

7.1 Conceptualization of Climate Security in EU Common Foreign and Security Policy

When comparing international initiatives and those developed by some EU member states to the addition of climate change to the EU security agenda, the latter has been slow to materialize. There has also been an evolution in the content of the climate security issue that began with the use of terms similar to those used in international practice, together with references to the risks faced by European territory. This evolution represents progress in the conceptualization of a notion of climate security for which there is no generally accepted definition. In international practice, a wide range of terms is used to define climate security. Climate change has been added to the security agendas in the form of such common expressions as risk, factor or multiplier of other threats to security. National strategies tend to include the basic idea that the effects of climate change can influence national security; however, from that point onwards it becomes difficult to establish a general terminology and characterization from the various definitions used by states to define their national security strategies.

An analysis of the evolution of the concept of climate security in the EU needs to begin with the 2003 EU Security Strategy. This text provided the conceptual framework for the EU Common Foreign and Security Policy and is particularly relevant in that it made no reference at all to climate change. By the time the Strategy was adopted, the EU had already set out relevant measures for the European space to mitigate climate change, including the Directive 2003/87/EC that set up a market for trading greenhouse gas emissions as the main instrument to fight climate change.

But above all, the EU had become the principal backer of the coming into force of the Kyoto Protocol. The abandonment of the Kyoto Protocol by the United States enabled the EU to develop its leadership role in international climate negotiations, which became a key objective in its foreign policy. Nevertheless, the first version of the 2003 Strategy did not include any link between this environmental question and foreign and security policy planning contained in it.

The main elements of reflection on climate security were added later, in the 2008 Report on Climate Change and International Security, presented to the Council of the EU by Javier Solana, High Representative for the Common Foreign and Security Policy (European Commission, 2008). By then, relevant data on the potential consequences of climate change had been made available by the IPCC; hence the report began by recognizing that "the risks posed by climate change are real and its impacts are already taking place". Meanwhile, the EU was preparing for the challenge of achieving the adoption of the second phase of compliance of the Kyoto Protocol at the meeting of the Parties to be held the following year in Copenhagen. With this aim in mind, the EU went to great lengths to design a negotiating strategy that would present European action as a model for confronting climate change.

In that context, the report analyzed the challenge of climate change and emphasized the need to take it into account when developing European foreign and security policy. Climate change is not defined in the text as an autonomous threat but as an enhancer of risks and threats in a complex international scenario. According to the text, "climate change is best viewed as a threat multiplier which exacerbates existing trends, tensions and instability" (European Commission, 2008). The transversal threats of climate change are cited as activators of conflict situations that were already considered dangerous. As previously mentioned, this approach coincided with actions being developed by some member states in their national security agendas, and also coincided with a debate at the United Nations Security Council at the behest, mainly, of the United Kingdom. Together with this, the report by EU High Representative Solana subscribed to the idea of linking climate change to the increase in threats to the maintenance of international peace and security and called for action to be taken based on this outlook.

Thus, a complex conceptualization of climate security developed, which would be inexorably subject to the case-by-case assessment of the array of factors that could influence a potentially vulnerable situation. The text points to certain "forms of conflicts driven by climate change which may occur in different regions of the world", including "conflict over resources", "economic damage and risk to coastal cities and critical infrastructure", "loss of territory and border disputes", "environmentally-induced migration", "situations of fragility and radicalization", "tension over energy supply" and "pressure on international governance". Although this list is by no means exhaustive, it does include various types of situations that would require a range of measures to be taken in response to them.

The Council of the European Union would take note of this report's conclusions, initiating a debate that would conclude with the definitive linking of climate change management to the development of foreign and security policy. As a result, the 2003 Strategy was revised to include climate change (Council of the European Union, 2008). This revision took place alongside the promotion of internal measures on climate and energy that would constitute the principal European position on climate change mitigation in international negotiations for the second phase of commitment to the Kyoto Protocol.

Interpreting this change yields contradictory results; in that climate change is a late addition to the EU security agenda, and its effect is considered only as a threat multiplier, while at the same time, the EU aims to convert its action on climate change into a transferable model in the advancement of the international judicial regime on climate change. This approach did not have the foresight to acknowledge the need for coherence and complementarity in the design of the EU Common Foreign and Security Policy, and in the environmental measures to mitigate, and adapt to, climate change, such that they could jointly strengthen the capacity to prevent and tackle conflicts, and provide the European territory with environmental protection. The need to develop this approach was recognized by the Council of the European Union itself in 2011 and, mindful of the negotiations to take place at the Conference of the Parties in Paris in 2015, acknowledged the need to revise the EU approach on climate change from the security perspective (Council of European Union, 2011).

A new phase in the evolution of the concept of climate security became apparent in *Shared Vision, Common Action: A Stronger Europe. A Global Strategy for the European Union's Foreign and Security Policy*, the document presented to the Council in June 2016 by Federica Mogherini, High Representative of the Union for Foreign Affairs and Security Policy, and Vice-President of the European Commission. The Global Strategy aimed to integrate the various objectives of EU action in the development of both internal policies and international initiatives. This text called for greater coherence in actions undertaken by the EU and its member states as a means to advance integration and to confront foreseeable crises. Prominence is given to concerns regarding the consequences of climate change and the need to consider their effects on foreign and security policy.

The Global Strategy does not provide an exhaustive list of the threats to be countered, nor does it clearly define what they are, meaning that the text contains no clear definition of what climate change is considered to be. Yet, the various sections dedicated to climate change show that there are two approaches to this environmental phenomenon and its relation to security. On the one hand, climate change is described as a threat with a multiplier effect on situations already considered hazardous, and this phenomenon "is a threat multiplier that catalyses water and food scarcity, pandemics and displacement". Elsewhere the text states that "climate change and environmental degradation exacerbate potential conflict, in light of their impact on desertification, land degradation, and water and food scarcity". The conceptualization of climate change is thus reiterated as a threat that is non-autonomous but one that aggravates already dangerous situations. In this aspect, it refers back to the approach adopted in the Solana report of 2008.

The Global Strategy document warns of the multiplier effect of threats from climate change in the sections of those threats that can arise in the international arena, and more precisely in the section on the objective to create "State and Societal Resilience to our East and South", which is deemed to be one of the priorities to be developed by EU foreign policy. By framing the notion of threat multiplier in these sections, the Global Strategy advocates the inclusion of climate security in any initiative aimed at preventing, and acting to counter conflicts, and therefore, it becomes a parameter to be considered in the development of European foreign and security policy within international peace and security actions. Thus it becomes an element in the conceptualization of climate security that influences EU foreign policy and is an acknowledgment that the effects of climate change will generate conflicts in the most vulnerable states, with international repercussions.

Furthermore, this conceptualization was complemented by the conceptualization of climate change as a risk to the people and territory of the EU, introducing the notion of concern over the possible adverse effects of this environmental phenomenon within the European space. It is specifically pointed out that "the EU Global Strategy starts at home. Our Union has enabled citizens to enjoy unprecedented security, democracy and prosperity. Yet today terrorism, hybrid threats, economic volatility, climate change and energy insecurity endanger our people and territory". In this second definition, climate change is now considered an autonomous risk, with specific influence on the internal area of the EU. Thus, there is mention of those forecasts of possible adverse effects of climate change which, as cited in the introduction, had been evaluated by the European Environment Agency, which includes rises in the sea level, the increased risk of adverse climatological events and natural catastrophes. This new approach is nothing more than an acknowledgement that climate change is on the march and accepts the reality of physical and social risks that need to be confronted with measures to assess their scope and planning.

This second approach to climate security figures in the section dedicated to "The Security of Our Union", which deals with the risks that would affect the EU territory, and is viewed as European security issues, but which are framed as foreign policy priorities. Yet beyond including climate change in this section, there is little to contribute to this definition of climate security. The lines of action listed in this section make no reference to climate change, however, there is a specific section on energy security, but this makes no mention of climate change either. So, although the need to secure energy supplies is given considerable importance, the issue of climate change is largely forgotten. The Global Strategy can be seen as a document that introduces the notion of concern about climate security to be included in programs but remains undefined in terms of lines of action, and it seems to be secondary to the securing of energy supplies, which are given a specific requirement.

The Global Strategy contains a concept of climate security that incorporates two different meanings of the connection between the development of climate change and the question of security. The first includes the conceptualization of climate change as a threat multiplier, the second, as a specific risk to the European population and territory. This enables the inclusion of the warnings cited by the IPCC and the European Environment Agency, and the description of the various vulnerabilities to climate change that territories and societies could be exposed to. With the acknowledgement of this diversity of climate security definitions, both in terms of confronting the effect of the threat multipliers implicit in climate change and in the prevention of, and action against, risk to European populations and territories, the Global Strategy proposes actions to be taken that are integral, internal and international in nature which take into account the need to adapt to, and mitigate, climate change. Within this framework, EU foreign and security policy should be integrated with the internal and foreign actions undertaken as part of all other EU policies, so that action on climate change and security policy become interconnected in a way that constitutes an integral idea of climate security.

The proposal to integrate climate change management with other EU actions in a coherent way fits with the general philosophy of the Global Strategy, which aims fundamentally for a new approach on European security that goes beyond merely enumerating the threats, and which insists on the need to develop a coherent European integration project. The Global Strategy underlines the need for planning general objectives, and the execution of those objectives by coordinated European action, and within the judicial system. In this context, climate security management becomes an element for consideration in this planned action, so that any preventive or reactive action that takes place in the face of its possible consequences is coherent. The challenge to transform this approach to climate security into specific measures with real environmental effect more or less connects to the general challenge of making a Global Strategy effective in order to override the internal distortions of a European system that make it difficult to coordinate the various European policies, with their internal and international aspects, with EU foreign and security policy. In the foreword by Federica Mogherini, she argues in favor of the Strategy in order to face down questioning of the objectives, and of the EU itself, by emphasizing the idea and the objective according to which "our Union needs a Strategy. We need a shared vision and common action".

So far, this study has traversed the introduction of the problem of climate change in the 2003 Security Strategy to the Global Strategy of 2016 to show how the concept of climate security in the EU has developed and how it has integrated into the EU's foreign and security policy. The study of this evolution would not be complete without referring to other strategic documents that incorporate climate change, and which give a broader perspective of the limits on forcing this concept onto the European agenda, such as documents that cover energy security and maritime strategy, for their close connection to climate change management both internally and internationally.

As already mentioned, the Global Strategy pays particular attention to energy security. Coherence between the measures adopted on energy security and climate security is vital for guaranteeing action on the environment that is sufficiently rigorous to mitigate climate change. An energy security policy based on achieving the objective of security of energy supplies but without provision for the environment can negatively affect any action on climate change. This makes it essential to include the objective of fighting climate change in planning for energy security. The *European Energy Security Strategy* adopted by the European Commission in 2014 includes the mandate to develop energy supply security in a way that is coherent and integrates with EU climate change policy (European Commission, 2014). This text emphasizes that energy supply objectives must be aligned with the regulatory initiatives included in the EU climate and energy package, thus acknowledging the need for transversal action on climate change. In particular, it states that "the Union's energy security is inseparable from the 2030 framework for climate and energy and should be agreed together by the European Council. The transition to a competitive, low-carbon economy will reduce the use of imported fossil fuels by moderating energy demand and exploiting renewable and other indigenous sources of energy".

The issue of maritime security is particularly relevant in terms of the treatment of climate change because the effects of climate change will affect maritime areas and coastal regions. The Council of the European Union adopted the *European Union Maritime Security Strategy* on 24 June 2014, to address the internal and external aspects of the EU maritime security

(European Council, 2014)[a]. The Strategy provided a list of the main maritime security interests of the EU and its member states, which included "the protection of the environment and the management of the impact of climate change in maritime areas and coastal regions, as well as the conservation and sustainable use of biodiversity to avoid future security risks". As with energy security, this document also recognizes the transversal nature of the effects of climate change which, in this case, would affect maritime areas and coastal regions. It thus reinforces the acknowledgement that climate security must include a specific definition related to the territory of the EU, as would later be included in the Global Strategy.

Specifically, the Maritime Security Strategy provided a definition of risk as the "potential security impact of natural or man-made disasters, extreme events and climate change on the maritime transport system and, in particular, on maritime infrastructure", adding that the response required consisted of "assessing the resilience of maritime transport infrastructure to natural and man-made disasters, including climate change, and to take appropriate adaptive actions and share best practices in order to mitigate related risks". Thus, a clear specific risk to the territory is cited in this document, which refers to significant social and economic impact, as well as issuing an order for preventive action to increase resilience to the effects of climate change. The next section features an analysis of the measures needed to manage this type of risk.

7.2 EU Action for Promoting Climate Security

As the narrative in the previous section testifies, the idea of climate security is now firmly on the European agenda, with the assumption that the effects of climate change represent significant risks and threats to security, and which require strategic planning and firm action to be taken. Nevertheless, the challenge ahead is for the strategies that have been designed to materialize in the form of specific actions. In other words, the objective is based on the programs described in the documents cited in the pre-

[a] *European Union Maritime Security Strategy*, adopted by the Council (General Affairs) on 24 June 2014, 11205/14.

vious section, to develop a regulatory system containing concrete binding measures. The practice of states and the actions developed within the United Nations framework underline the difficulties in moving forward and beyond the security agenda declarations on the interconnection between climate change and security in order to achieve clarity on the binding measures required to mitigate, and adapt to, climate change. The Global Strategy text alluded in general terms to the difficulty it foresaw in reaching its objectives, even going so far as to state explicitly that the measures to be implemented would require a transition "from vision to action". The Global Strategy presents an approach and objective that are ambitious in their attempt to incorporate into European policies an area as complex as the development of the EU Common Security and Defence Policy, to the extent that the introduction of evidently non-military elements such as climate change amounts to a challenge that is considerable and complex (Smith, 2017).

When addressing how to put its objectives into practice, the Global Strategy sometimes refers specifically to climate change, and even though it does not refer directly to this phenomenon, the actions it proposes are largely seen as for application in this area. As previously mentioned, the cornerstone of the Global Strategy is the need to coordinate the objectives and development of foreign and security policy across all the other European policies, hence the emphasis on the guiding principle of "unity in action", both by applying coherent policies and in the relationship between the EU and its member states. The Global Strategy states that "the people of Europe need unity of purpose among our Member States, and unity in action across our policies"; and "the interests of our citizens are best served through a unity of purpose between Member States and across institutions, and unity in action by implementing together coherent policies". In line with this is the idea that action on climate change becomes one of those parameters that fits the notion of unity in action because its transversality marks it out as an element with the potential to integrate European action. This approach implies the need for more interconnected action in environmental and security matters because implementing climate security means rethinking the EU's foreign and security policy from an environmental perspective. Thus, the complexity and challenge of making this vision and programme on climate change a reality and putting it into practice are linked to the capacity of the Strategy to be effective. For example, Youngs sounded the alert on the setting up of numerous departments that act without any connection between them, far removed from the necessary strategic organization that action based on climate security required (Youngs, 2014).

In terms of shaping the external dimension of climate security by addressing the multiplier effect of threats to climate change and the internal dimension in relation to the risks of the effects of climate change on populations and territories, the Global Strategy describes some of the specific challenges that the actions it proposes need to address. Stating that climate change is a multiplier of potential threats emphasizes the need to integrate the concept of climate security into the development of foreign policy as a whole. This idea leads on to the demand to play a leading role in the international legal regime on climate change, at a time like the present when uncertainties over the future of climate control negotiations are more evident than ever. The latest development of the international legal regime on climate change resulted in the adoption of the Paris Agreement, but US President Donald Trump announced at the beginning of June 2017 his intention to withdraw the United States from the treaty. All the documents produced by the UN Security Council, UN General Assembly or the UN Secretary General recognize that the international regime on climate change is the ideal forum for debating issues related to this environmental phenomenon, so support for the continuity and efficacy of this forum for negotiation needs to be strong and consistent. In agreement with this approach, several Western states and emerging nations have expressed their intention to continue with the Paris Agreement project despite the fact that the United States withdrawal opens up a new phase of uncertainty in the negotiations on climate change.

The EU has traditionally been one of the guarantors of the continuity of this international regime and has played an important leadership role in doing so, although it has not always achieved its objectives. There is now a new opportunity to develop a position of leadership in climate change matters that requires it to consolidate relationships with non-EU states that back the international judicial regime on climate change and to maintain an active presence in all the international regimes involved in the complex global governance of climate change. In consequence, the EU has openly supported the Paris Agreement, and this is clearly stated in the first report on the implementation of the Global Strategy (European External Action Service, 2017).

In other words, the EU needs to develop a style of leadership within the various sectors of global climate governance. This fits with the general calls in the Global Strategy to develop a leadership role in international negotiations, to develop a dialogue with potential partners with similar interests and to strengthen the framework of the United Nations system of negotiation, all of which are challenges to be taken into account when considering action in the area of climate change.

Once again we find ourselves in awkward and controversial territory. The new international scene hinders this desire for leadership, namely in terms of the new position adopted by the United States that is contrary to deepening climate change commitments, and the position of states such as China which aims to be a protagonist in climate control negotiations. Traditionally, the EU has based its leadership in climate issues on the credibility of its position sustained by the development of an internal regulatory model for action to counter climate change that is both substantial and effective, but this style of leadership has its limitations. The idea of leadership by example has been widely debated by the internationalist doctrine, which has indicated both its great potential and its objections (Bäckstrand and Elgström, 2013; Gupta and Grubb 2000; Oberthür and Groen 2017; Parker and Karlsson 2010; and Wurzel and Connelly 2011).

The counterproductive effects in international negotiations of the EU's excessive tendency to try to get other states to adopt regulations that it has developed within the European setting have already been noted (Youngs, 2014). This is especially evident in the market for greenhouse gas emissions trading which has been strongly defended by the EU. There is no doubt that the development of certain European policies on climate change is due to the concurrence of factors with their roots in the European setting, not to mention institutional factors, and which are hard to transfer to other national or international situations (Mehling et al. 2013). The EU's leadership on climate change must, therefore, get the balance right in terms of the importance it gives to defending its regulatory model to confront climate change. Although this model is positive in terms of generating confidence because it shows the importance of sticking to commitments made, it can also be seen in negotiations on climate change as an imposition of rigid parameters.

And although the question of European leadership in climate governance has been widely explored and debated, there have been few developments on the integration of the concept of climate security in the institutional framework as part of the European policy framework on climate security and defense, and EU participation in crisis management missions. Little has been done to incorporate climate factors in conflict prevention policies; thus the potential for prevention is wasted when the response to a conflict is proposed (Youngs, 2015). And this despite the fact that civilian and military missions could be a good scenario for the design of policies to mitigate, and especially adapt to, climate change, action in these areas would require the use of specialized methodologies and, hence, a specific environmental capability to develop operations from which this component has traditionally been absent. Despite the obvious difficulties in developing action in this scenario, it represents a challenge that fits with calls from the Global Strategy to strengthen the resilience of states, and which can reinforce the maintenance of peace and security.

As indicated in the previous section, the Global Strategy also defines climate change as a risk to people and territory within the EU, thereby endowing climate security with significance for the interior of the European territory itself. Addressing this risk requires adopting ambitious actions to mitigate, and adapt to, climate change. Specifically, the articulation of the concept of climate security can be of particular interest for its interconnection with the development of measures for adaptation within the European territory. Identifying a risk to people and to the territory within Europe itself highlights the need to develop adaptation measures in this scenario as well, which must necessarily include measures for any rapid response to an emergency. The EU's climate change policy in recent decades has emphasized the development of legal measures for climate change mitigation, systematized within the climate and energy package. However, the reality of the advance of climate change that is now happening shows the need to bring adaptation forward, and the proximity of its effects from the security perspective represents a major incentive for the development of ambitious binding commitments.

As already stated, in 2013 the European Commission adopted the *European Union Strategy on Adaptation to Climate Change*, which established the general framework for developing legal action for adaptation within the European system. The EU Adaptation Strategy assumed the need for action across sectors and levels to respond to the complex challenge of the multiple impacts of climate change on European territory. The Strategy included three types of action to be taken; first, to foment actions for adaptation at the national level, second, action on adaptation by the EU in vulnerable sectors, and third, the development of a European system of information on adaptation. These actions were aimed at achieving greater resilience across the European territory to the adverse effects of climate change.

The development of the Adaptation Strategy, therefore, represents a fundamental element in guaranteeing climate security. In its opening pages, the Strategy unveils an explanation of the current and future effects of climate change that cover environmental, social and security aspects, and includes a call for action without further delay. In particular, it warns that failure to act, or any delay in acting, could undermine the cohesion of the EU, thus determining climate change as a risk to the process of European construction itself. At present, it is necessary to develop the Strategy's various fields of action; and it could also be argued that the implementation in practice of climate security requires approval of a plan for adaptation to climate change in Europe in the very near future that goes beyond the Adaptation Strategy, and which is a real plan for adaptation to climate change on the European level. There is no doubt that the EU faces a considerable challenge in this field in which the security-based approach adds a new element to incentivize the adoption of measures that have previously been hard to negotiate from a purely environmental perspective. The security-based approach to climate change focuses on the risks and threats from this environmental phenomenon, and this could be an interesting incentive to include ambitious objectives in decision-making on the environment, and which contributes new operational systems for prevention, planning and action.

In the short term, the approach of adaptation to climate change will undoubtedly include greater incentives to adopt that measures in the Adaptation Strategy that relate to vulnerable sectors. One example could be the determination of specific risk to maritime routes and infrastructure that figures in the Maritime Security Strategy referred previously. This Strategy calls for the adoption of adaptation measures that promote resilience to climate change. The scope of the EU's integrated maritime policy thus becomes a legal area of particular relevance in the development of adaptation measures to counter the risks described in the Maritime Security Strategy. Once again, the need for coherence and integration of the policies defended in the Global Strategy becomes vital for the development of specific legal measures that transform the idea of climate security into practical actions.

Finally, there is a special mention of the need to integrate the requirements of energy security with climate security. The fact that energy security and climate security can be complementary has already been discussed, with regard to the objective of seeking an energy transition that promotes the use of energy sources with low impact of greenhouse gas emissions. However, the aim of securing self-sufficiency in energy supplies alone could generate distortions that might render the effects of action in one field contradictory to action in another. In reality, this means integrating the objectives of the EU's Energy Policy as set out in Article 194 of the Treaty on the Functioning of the European Union. This precept emphasizes the security of energy supplies and urges greater interconnection of energy networks but also calls for energy-saving measures and the development of renewable energies. Article 194 requires that actions across various sectors be integrated, because interconnection is linked to foreign policy and common security, and the latter is interrelated with a competitive and sustainable economy. As indicated previously, the *European Energy Security Strategy* emphasizes this connection, and it will be in the development of this connection that the requirements for achieving a coherent energy supply will have to integrate with measures to mitigate and adapt to climate change.

Conclusion

The EU has determined to make its action on climate change one of the mainstays of its activity within EU territory and internationally. This vision is defined by the development of a regulatory system, in which action to mitigate, and adapt to, climate change has been formulated as a set of legal transversal and sectoral measures with specific aims and deadlines; these measures are now integrated in a broader series of actions for achieving a system of development that is both competitive and sustainable, as well as in foreign policy, at a time when climate diplomacy is increasingly important as a means to maintaining a credible presence in the main climate negotiation forms.

It is within this framework that the conceptualization of, and action on, climate security has established itself, and where this approach has the capacity to generate greater incentives to achieving environmental goals and harmonization across different sectors. The idea of climate security comprises both a securitization process in the climate change management approach and introduces the component of environmental transversality in foreign and security policy.

The EU's Global Strategy is the product of experiences from previous practice, and definitively establishes climate security as a parameter of foreign and security policy. The Global Strategy promotes the strategic coherence and connection of the climate security objective to EU policies for action both inside and outside the European territory and to actions undertaken by its member states. The security-based approach is an incentive to develop coordinated action. The transversality of the climate element could incentivize coordinated decision-making that takes into account general objectives and principles for action, with better regulation being the ever-present aspiration, in this case within a broader context in which legislative activity within the EU interrelates with the development of EU foreign policy. The challenge for climate security is thus similar to the challenge of effective implementation of the Strategy, in that there is a need for planned and coordinated action globally that can yield coherent results in the actions carried out by the EU and its member states.

The challenge of transforming the idea of climate security into specific measures is not easy precisely because it reflects deficiencies in the development of European foreign and security policy and even those relating to the integration process itself. The roll-out of its objectives requires the EU to interconnect the legislative process with its actions, which would amount to a substantial change in the system of European integration. The idea of climate security thus becomes one more element in the EU's determination to achieve internal development and international leadership based on the defense of the struggle against climate change, and in the promotion of greater resilience in the face of its adverse consequences. Action on climate change has become a badge of identity for the EU that is displayed in the international arena as a model for sustainable development; action on security will present the opportunity to develop a new model of adaptation to climate change and its most violent effects.

References

Adger, W.N. et al., (2014) AR 5 climate change 2014: Impacts, adaptation, and vulnerability, in C. B. Field, et al. (eds.), *Human Security.* Cambridge University Press, Cambridge, pp. 755–791.

Bäckstrand, K. and Elgström, O. (2013) The EU's role in climate change negotiations: From leader to "Leadiator." *Journal of European Public Policy*, 20(10), 1369–1386.

Balzacq, T. (2010) *Securitization Theory: How Security Problems Emerge and Dissolve.* Routledge, London.

Brauch, G., Link, P.M. and Schilling, J. (eds.) (2012) *Climate Change, Human Security and Violent Conflict. Challenges for Societal Stability*, Springer, Berlin.

Buzan, B., Wæver, O. and de Wilde, J. (1998) *Security. A New Framework for Analysis*, Lynne Rienner Publishers, Boulder.

Council of the European Union (11 December 2008) *Report on the Implementation of the European Security Strategy: Providing Security in a Changing World*, S407/08.

Council of the European Union (18 July 2011) 3106th Foreign Affairs Council meeting, *Council Conclusions on EU Climate Diplomacy.*

EEAS (2016) *Shared Vision, Common Action: A Stronger Europe. A Global Strategy for the European Union's Foreign and Security Policy*. European External Action Service, Brussels.

European Commission (2013) *An EU Strategy on adaptation to climate change.* Communication from the Commission to the European Parliament, the Council, the European Economic and Social Committee and the Committee of the Regions, 16 April 2013, COM(2013) 216 final.

European Commission (14 March 2008) *Climate Change and International Security.* Paper from the High Representative and the European Commission to the European Council, S113/08.

European Commission (2014) *European Energy Security Strategy*, Communication from the Commission to the European Parliament and the Council, 28 May 2014, COM (2014) 330 final.

European Commission Communication (3 March 2010) *Europe 2020: A Strategy for Smart, Sustainable and Inclusive Growth*, [COM (2010) 2020].

European Council (2014) *European Union Maritime Security Strategy*, adopted by the Council (General Affairs) on 24 June 2014, 11205/14.

European Environment Agency (2017) *Climate Change, Impacts and Vulnerability in Europe 2016. An Indicator-Based Report*, Publications Office of the European Union, Brussels.

European External Action Service (2017) *Implementing the EU Global Strategy. Year 1 Report*, available at: http://europa.eu/globalstrategy/

Gupta, J. and Grubb, M.J. (eds.) (2000) *Climate Change and European Leadership. A Sustainable Role for Europe?*, Kluwer Academic Publishers, Dordrecht.

Harris, K. (2012) *Climate Change in UK Security Policy: Implications for development Assistance?*, Overseas Development Institute, London.

High Representative of the Union Foreign Affairs and Security Policy (2016) *Shared Vision, Common Action: A Stronger Europe. A Global Strategy for the European Union's Foreign and Security Policy*, available at: http://eeas.europa.eu/archives/docs/top_stories/pdf/eugs_review_web.pdf

Mehling, M., Kulovesi, K., and de Cedra, J. (2013) Climate change and the law, in Hollo, E.J, Kulovesi, K. and Mehling, M. (eds.), *Climate Law and Policy in the European Union: Accidental Success or Deliberate Leadership?*, pp. 509–522.

Oberthür, S. and Groen, L. (2017) The European Union and the Paris Agreement: Leader, mediator, or bystander?. *WIREs Climate Change*, vol. 8, pp. 1-8.

Page, E. (2000) Theorizing the link between environmental change and security. *Review of European Community & International Environmental Law*, 9, 33–43.

Parker, C.F. and Karlsson, C. (2010) Climate change and the European Union's leadership moment: An inconvenient truth? *Journal of Common Market Studies*, 48(4), 923–943.

Scott, S.V. (2012) The securitization of climate change in World politics: How close have we come and would full securitization enhance the efficacy of global climate change policy? *Review of European Community & International Environmental Law*, 21, 220–230.

Security Council (17 April 2007) *Security Council Holds First-ever Debate on Impact of Climate Change on Peace*, SC/9000.

Smith, M.E. (2017) *Europe's Common Security and Defence Policy. Capacity-Building, Experiential Learning, and Institutional Change*, Cambridge University Press, Cambridge.

United Nations Security Council (2011) *Statement by the President of the Security Council*, 20 July 2011 (S/PRST/2011/15).
Wurzel, R. and Connelly, J. (eds.) (2011) *The European Union as a Leader in International Climate Change Politics*, Routledge, London.
Youngs, R. (2014) *Climate Change and EU Security Policy. An Unmet Challenge*, Carnegie Endowment for International Peace, Brussels.
Youngs, R. (2015) *Climate Change and European Security*, Routledge, London.

Legislation

Directive 2003/87/EC of the European Parliament and of the Council of 13 October 2003, establishing a scheme for greenhouse gas emission allowance trading within the Community and amending Council Directive 96/61/EC (OJEU L 275, 25 October 2003).
Kyoto Protocol to the United Nations Framework Convention on Climate Change, adopted in Kyoto, Japan, on 11 December 1997, which entered into force on 16 February 2005. *UN Treaty Series*, A-30822.
Paris Agreement at the 17th Conference of the Parties to the United Nations Framework Convention on Climate Change, 30 November to 11 December 2015. See Decision 1/CP.21, *Adoption of the Paris Agreement*, available at: http://unfccc.int/2860.php

CHAPTER

8

The Role of European Cities in the Fight against Climate Change: Networking for Local Climate Policy Convergence

Ekaterina Domorenok

Cities can be the engine of social equity and economic opportunity. They can help us reduce our carbon footprint and protect the global environment. That is why it is so important that we work together to build the capacity of mayors and all those concerned in planning and running sustainable cities.

–Ban Ki-moon (2012)

Introduction

Building resilience and contributing to the global fight against climate change have been among the top European Union's (EU's) political priorities over the last decade. The urban dimension has received particular attention in this context in view of the fact that urban energy consumption generates about three quarters of global carbon emissions (European Commission, 2015), whereas local governments' jurisdictions widely cover most of policy sectors relevant for tackling this negative trend. Moreover, local authorities play a crucial role in engaging local communities and stakeholders into governing processes at the local level and can effectively enhance more sustainable ways of living. Consequently, cities have been increasingly recognised as core actors of the EU energy and climate governance, as they could enable a shift in energy production and consumption to more sustainable pathways, contributing also to the creation of new opportunities for investments and jobs (JRC, 2016).

A range of policy initiatives have been promoted by the EU over the last decade to strengthen the role of cities for sustainable development and, at a later stage, in climate policies. In particular, horizontal cooperation and networking among cities have been considered to be an important tool for boosting their potential in terms of policy innovation and learning for climate, as it stems from a genuine bottom-up commitment to and strong ownership of common policy goals. Thus, in addition to the most well-known URBACT programme launched within the EU cohesion policy during the 2007–2013 programming period to support cities in designing and upgrading integrated policies for sustainable urban development, a range of smaller measures targeting cities and their collaborative partnerships have been implemented within the framework of environmental, innovation and development policies. These include voluntary initiatives, awards and pilot projects funded through the Life+ or Horizon 2020 programmes, to mention a few. Drawing on such evidence, it has been suggested that the implementation of EU climate change policy at local level alters local practices and policies and, vice versa, local climate governance influences the development of EU climate policy (Kern, 2010; Kern and Bulkeley, 2009).

This chapter analyses the experience of the Covenant of Mayors (CoMs) programme that was launched by the EU Commission in 2008 to enhance and support local action for climate change across EU countries. To join the programme, municipalities are required to voluntary commit to the EU Energy and Climate Package targets and make efforts to prepare local Sustainable Energy and Climate Action Plans (SECAPs) that substituted Sustainable Energy Action Plans (SEAPs) in 2015. Such plans should include a set of measures for the reduction of CO_2 emissions from a wide range of sectors not covered by the Emissions Trading System (ETS), including public and residential buildings, local productive sector, transport, waste and lightening. Along with specific methodological guidance for the preparation of SE(C)APs and a scheme for conducting regular monitoring and benchmarking, a range of knowledge resources and capacity-building facilities have been set out by the programme to encourage coordination and learning among its signatories. Other territorial entities (regions and provinces) and specialised agencies have also been encouraged to join and support municipal actions towards climate change with their financial, networking and knowledge resources.

Such a complex multilevel programme design has aimed at mobilising a multitude of public and private actors in the effort to combat climate change within and beyond EU borders, as the participation in the initiative was opened also to cities from extra-EU countries. While performing typical functions of Transnational Municipal Networks (TMNs) such as information sharing, capacity building and rule setting, the CoM distinguishes itself substantially from these networks in terms of the nature of signatories' commitment, the composition of partnership and the organisational structure (Bendlin, 2015). More specifically, the political objectives and targets established by the CoM has been shaped by the EU 2020 Energy and Climate Package (European Commission, 2007), while its coordination has been carried out by the Covenant of Mayors Office (CoMO) that is in charge of the overall functioning of the programme and relies on financial support of the EU Commission. The EU Joint Research Centre (JRC) offers technical and expert assistance required for designing and implementing local action plans.

In addition to presenting an overview of the Covenant's evolution over time, this chapter describes the functioning and the main policy tools of the programme to enhance policy and governance change at the local level. It also provides some evidence of how and why cities join this initiative and reflects on opportunities and commitments deriving from the CoM membership.

8.1 The Establishment and Evolution of the CoM: Work in Progress

As is known, the EU regulatory package adopted to comply with international commitments within the United Nations Framework Convention on Climate Change (UNFCCC) has been laid down by two main pillars: the ETS established by the EU Directive 2003/87/EC (European Union, 2003) and the Effort Sharing Decision (Decision No 406/2009/EC) adopted in 2009 (European Council, 2009). While the aforementioned provisions have mainly targeted Member States, the role of local authorities has been clearly recognised and increasingly enhanced by subsequent policy initiatives, including the Energy Efficiency Directive (European Union,

2009), the European Energy Union Package (European Commission, 2015), the European Heating and Cooling Strategy (European Commission, 2016a) and the European Strategy for Low-Emission Mobility (European Commission, 2016b).

The CoM was launched by the EU Commission to enhance the low carbon efforts of European towns and cities within the framework of the EU 2020 Energy and Climate Package (European Union, 2008), particularly in the policy sectors covered by the Effort Sharing decision. The purpose of the EU's policy commitment has subsequently been upgraded and a new initiative—Mayors Adapt—was launched in 2014 to further increase local efforts for climate by adding the adaptation objective to the already developed mitigation agenda. Thus, the Mayors Adapt initiative has encouraged local governments to develop new strategies or upgrade old ones to embrace new mitigation targets going beyond the energy efficiency and renewable energy objectives sources to wider climate adaptation goals. The two initiatives worked in parallel for 1 year before being merged in a unique programme entitled the Covenant of Mayors for Climate and Energy. The feasibility of such transition has been discussed and welcomed as a result of an extensive consultation process among CoM participants, which helped shape the new set of goals and commitments of the programme. Climate adaptation objective has been added to the programme's mitigation targets, aiming to achieve 30% reduction in CO_2 emissions by 2030. Thus, currently, CoM signatories pledge to actively support the implementation of the EU mitigation and adaptation targets by 2030 and commit to adopt an integrated approach to climate policies, while ensuring access to secure, sustainable and affordable energy for all. This new long-term vision embraces the following three broad strands of action:

- Accelerating the decarbonisation of the territories of its signatories, thereby maintaining the average global temperature rise below 2°C
- Strengthening the capacity of its signatories to adapt to unavoidable climate change impacts to make their territories more resilient
- Increasing energy efficiency and the use of renewables on the territories of its signatories to ensure access to secure, sustainable and affordable energy services for all

The scope of the CoM has rapidly expanded as well. In 2010, the programme enrolled about 2,000 signatories and it doubled its membership totalling 4,330 signatories in 2013. As Table 8.1 shows, as of January 2019 the CoM comprised 7,383 signatories from EU countries, 270 signatories from the Eastern Partnership and 102 local authorities from other countries. The scope of the CoM membership has extended eastwards through the Covenant of Mayors East initiative that started in 2011, involving Belarus, Ukraine, Moldova, Armenia, Georgia and Azerbaijan. In 2012, the Cleaner Energy-Saving Mediterranean Cities (CES-MED) project was launched to enlarge the Covenant to the European Neighbourhood South Region covering cities in Algeria, Egypt, Israel, Jordan, Lebanon, Morocco, Palestine and Tunisia. Later new regional offices were established in North and South America, Japan, India, China and south-east Asia.

In 2017, the Global Covenant of Mayors for Climate and Energy (GCoM) was launched with the aim to join together the European and global efforts of local authorities, by bridging the EU's CoMs and the Compact of Mayors. The new partnership has been expected to strengthen the crosscutting links between the CoM and TMNs for climate, such as ICLEI, C40 and Climate Alliance, with which it has developed multiple crosscutting linkages since its very inception (Busch, 2015). The GCoM has become the largest network of local authorities committed to the United Nation's efforts for climate, whose objectives have recently been updated by the Paris agreement. As stated by its mission, this network aims to 'accelerate ambitious, measurable and planed climate and energy action that lead to an inclusive, just, low-emission and climate-resilient future, helping to meet

Table 8.1 CoM signatories as of January 2019.

Region	Number of Signatories	Inhabitants	Coordinators	Supporters
EU – 28	7,383	198,038,338	185	161
Eastern Partnership	270	22,850,513	22	13
Europe – non EU	47	5,557,452	0	6
South Mediterranean	36	4,571,477	0	0
Others	19	21,610,988	1	0

Source: Author elaboration on the CoM data available at: https://www.covenantofmayors.eu/

and exceed the Paris agreement objectives' (GCoM). City and regional networks, national governments and other partners are welcome to join.

The overall rise of transnational climate city initiatives can be explained by the increased awareness about the need of joint action (Jordan et al., 2015) and the demand for expert knowledge required for developing effective local climate policies (Jordan et al., 2018). However, the relevance of city networks appears to be ultimately determined by their actual capacity to coordinate joint efforts and shape the process of policy change at the local level. The CoM has developed a range of methodological tools to this end, which local authorities are expected to adopt to converge their policies towards shared guiding principles and operational patterns. Indeed, given the CoM's deliberate objective to persistently shape and support the local efforts in the implementation of sustainable energy and climate policies in the long-term perspective, it has been suggested that differently from other TMNs the CoM can be conceived as 'transnational environmental regulation regime' that is able to drive policy and governance change for climate (Heyvaert, 2013).

8.2 Policy Convergence through Coordination: Guiding Principles and Operational Tools

The possibility to sign up to the Covenant is open to any municipality willing to commit to the programme objectives and targets. There is no specific deadline for joining the CoM and the membership is free of charge. The coordination function and members' operational assistance are performed by the CoMO, while the JRC provides the required technical guidelines, methodological documents and templates to assist municipalities in the preparation and implementation of their local action plans.

Regardless of the voluntary nature of the CoM, a number of specific rules and procedures, which local authorities must comply with, have been established to become and remain part of the network. Such rule-setting framework has been aimed at both encouraging local authorities to establish common carbon emission targets and coordinating their action by adjusting local policies to shared principles and methodologies.

As far as the general political ambition is concerned, the transition from the CoM for sustainable energy to the Covenant for Climate and Energy in 2015 has resulted in an important upgrade of the programme's priorities, although opening the path for a greater variation of participants' commitment. For those who joined the programme before 2016, the commitment remained framed by the 2020 Energy Package as defined by the initial Commitment Text, according to which CoM signatories must submit a SEAP within 1 year after their formal signing to the programme, committing to reduce their emissions beyond the 20% European emission reduction target by 2020, while simultaneously increasing energy efficiency and the share of energy produced by renewables.

Since 2016, in addition to mitigation targets, CoM signatories have been invited to develop policies for resilience to climate change, as well as intensify their efforts in providing access to secure, sustainable and affordable energy for all. Therefore, they have been required to develop a local adaptation plan or a strategy and/or mainstreaming adaptation into existing relevant plans within 2 years of signing up to the initiative. All signatories working within the 2020 mitigation objective and those committed to the adaptation agenda were encouraged to renew and upgrade their commitments to the 2030 targets with a new Municipal Council deliberation and a 2030 adhesion form. For those signatories who joined the programme after 2016, a target of 40% emission reduction by 2030 was established.

Local action plans are subject for approval by the JRC and the monitoring procedure should be set out within 2 years from the approval date. The implementation progress should be evaluated based on the collected monitoring data to revise and adjust policy objectives and measures. The so-called 'full monitoring' should be carried out within 2 years after the first step 'action' monitoring is performed. In addition to the assessment of the effectiveness of implemented measures, this type of monitoring requires also a comprehensive revision of Basic Emission Inventories (BEIs). In other words, a step-by-step process has been designed for local authorities to follow when developing and implementing sustainable energy and climate policies included in their local plans. All these steps have been supported by methodological and capacity-building tools developed by the CoMO and the JRC.

In this way, the process starts with the creation of BEIs that collects as much as possible the comprehensive data on the main sources of CO_2 emissions in a given territory, for which the respective reduction potential should be clearly identified. A template and the related technical guidance have been provided for designing BEIs that represent a first fundamental building block for developing SE(C)APs. Following the launch of the Covenant for Climate and Energy, local authorities have been asked to include the assessment of the main climate risks, vulnerabilities and future challenges in their preparatory analysis.

It is worth stressing that although CoM templates trace the overall methodological guidance, their layout is flexible enough to allow for adjustment to very different context conditions, including sectors to be addressed, emission factors to be considered and emission units to be used for reporting. In fact, although a general recommendation was made to use 1990 as the year for the BEI reference, signatories could choose the closest subsequent year for which reliable data could be gathered. As a result, different years of reference have been chosen for constructing BEIs and the list of priorities to tackle identified by local plans varies substantially across CoM members. The primary potential impact of the CoM action in term of CO2 reduction was expected to concentrate in the following sectors: municipal buildings, equipment and facilities; tertiary buildings, equipment and facilities; residential buildings; public lighting; industry; transport; local electricity production and local heat/cold production (Croci et al., 2016).

As far as the climate adaptation strand of action is concerned, a specific methodology has been developed, allowing municipalities to keep their framework of reference sufficiently open to include the evolving conditions of signatories. This framework builds on the so-called Risk and Vulnerability Assessment (RVA) approach, which should be integrated into the SECAP and mainstreamed through other relevant planning documents. The RVA aims to develop a comprehensive picture of current and future climate change impacts, particularly in urban areas, by identifying potential risk and stress factors, as well as opportunities arising from climate shifts. It also provides information on how to assess the local adaptive capacity to cope with the expected adverse impacts and understand how climate change affects social and economic dimensions.

The methodology underlying the RVA combines the top-down and bottom-up approaches and suggests to use quantitative and qualitative data, aiming to cope with the problem of uncertainty and knowledge gaps in the perspective of climate adaptation. In this way, a map should be developed on how a range of climate factors are likely to change in a specific area, how people's life is likely to be affected by these changes and how impacts may vary across sectors. Municipalities should also specify the year when the RVA was carried out, as well as the area of reference (e.g., municipal, metropolitan, provincial, regional) and the method used. The aforementioned methodological elements have laid down the basis for the so-called Urban Adaptation Support Tool (UAST), which is currently widely promoted for developing mitigation strategies. This tool departs from the assessment of risks and vulnerabilities, and proceeds with the identification of adaptation options in the perspective of their future implementation, monitoring and evaluation. Within this framework, most common hazard types at urban level are mapped and adaptation indicators are identified.

The compliance of local action plans with the CoM methodological and technical guidance adopted is checked by the EU JRC that approves plans based on a set of eligibility criteria, including the consistency of the collected data, coherence of local targets with EU objectives, and the scope and appropriateness of measures (CoM Reporting Guidelines, p. 3).

Along with individual commitments, the CoM has envisaged the possibility to develop collective action plans, which has been particularly relevant for small neighbouring municipalities. Within this framework, each municipality should sign up choosing between the following two options:

- *Individual commitments*: Each signatory individually commits to reducing CO_2 emissions of 20% by 2020 and/or 40% by 2030, completing its own template with an action plan that can contain both individual and shared measures.
- *Shared commitments*: The group collectively commits to reducing CO_2 by 20% and/or 40%. In this case, only one single template is to be filled in by the entire group, which is then listed under a grouped signatory.

As for monitoring, the JRC has developed and updated the dedicated template as well, considering the need to integrate mitigation and adaptation actions into local strategies. The current monitoring framework is composed of a scoreboard and tables with pre-defined cells, covering measures and actions as well as resources and barriers that might be encountered during implementation. Such a framework aims to homogenise reporting criteria across countries, encouraging local authorities to refer to the same standards and methodology that should be necessarily adjusted to their individual conditions. Yet, the monitoring effort of CoM signatories has been limited so far, as in 2016 only 1743 municipalities submitted their action monitoring reports, and only 315 signatories (6% of the total) reported on the implementation of their action plans by presenting the so-called full monitoring (JRC, 2016, p. 28).

When implementing SE(C)APs, local authorities have been invited to share successful practices through the dedicated section in the CoM website to provide a benchmark for all those interested in improving their actions by borrowing from others' experience.

The whole SECAP cycle is summarised by Table 8.2.

Table 8.2 SECAP step-by-step process.

Steps	Actions	
1. Initiation and baseline review	Preparing a **Baseline Emission Inventory (BEI).**	Preparing a **Climate Change Risk and Vulnerability Assessment.**
2. Strategic target setting and policy planning	Drawing a **Sustainable Energy and Climate Action Plan (SECAP)** and designing mitigation and adaptation measures across relevant local policies, strategies and plans (within 2 years from the Municipal Council decision to sign up).	
3. Implementation, monitoring and reporting	**Reporting** progress every second year following the SECAP submission in the CoM platform.	

Source: Author's elaboration on the CoM Commitment Document, available at: https://www.covenantofmayors.eu/support/library.html

In addition to the regulatory guidance and capacity-building tools described above, several training and information events have been promoted by both the CoMO and individual participants with the purpose of creating and spreading general and specific knowledge on climate and energy policies at the local level. However, as the paragraph below illustrates, the local response to the aforementioned opportunities has been disproportionate.

8.3 Opportunities and Costs of the CoM Membership

Although the overall purpose of the CoM is clearly defined, municipalities may have different motivations for joining the programme, and their perceptions about the programme's effectiveness vary considerably. Studies[a] conducted to understand why municipalities choose to participate and to what extent their expectations have been met show a great variability of perceptions on the potential and actual opportunities of the CoM, as well as on participation costs and benefits.

Thus, according to a survey administered in 2013 (Eparvier et al., 2013), the objective to demonstrate political commitment to climate mitigation agenda has been by far the most important motivation for municipalities to participate, followed by the expectation to gain access to technical knowledge and EU funding. Opportunities such as job creation and economic growth, coordination with other municipalities and the chance to obtain broader technical/scientific support for climate policies have been perceived as less relevant. The possibility to acquire visibility at the EU, national and regional levels, has been viewed as the least important aspect.

Overall, the opinions about the actual added value of the CoM show that municipalities' expectations have largely been met. Among the most important benefits of the CoM, the following aspects have been mentioned[b] (Eparvier et al., 2013):

- Access to knowledge (88%)
- Better access to EU funding (83%)
- Technical guidance and support (83%)
- Increased mobilisation of stakeholders (82%)
- Increased support from region (76%)

---

[a]Among others these include the following: JRC (2016). Covenant of Mayors: Greenhouse Gas Emissions. Achievements and Projections; Technopolis Group et al. (2013). Mid-term evaluation of the Covenant of Mayors, Final Report; CoM (2017). Covenant community's need for SE(C)AP design and implementation. The general limit of these studies lies in the fact that very active and engaged signatories may be overrepresented in the sample of respondents.

[b]The sample of 553 municipalities has been considered by the Report Mid-term Evaluation of the Covenant of Mayors. Final report, 2013 (by Technopolis Group, Fondazione Enrico Mattei, Hinicio, Ludwig Bölkow Systemtechnik).

- Enhanced visibility towards national governments, European Commission and the other signatories (73%)

The above data and related comments from interviews show an important impact of the CoM as a tool for the diffusion of knowledge and technical assistance in the field of sustainable energy and climate (Eparvier et al., 2013). The programme has proved to be relevant for enhancing local capacities to design and implement sustainable energy action plans, as 63% of respondents considered that the CoM was effective in the following core aspects of local energy and climate planning:

- Designing a mix of mitigation policies and actions
- Increasing the coherence of actions already designed in the field of energy efficiency and renewable energy production
- Mainstreaming actions for climate across sectorial policies

Municipalities have particularly appreciated the possibility to learn about technical solutions to reduce CO_2 emissions by developing collaboration with universities and private companies. Overall, differences in the perceptions on the aforementioned issues appear to be partly determined by national contexts, as well as by the level of expertise and technical capacities of individual municipalities, which often depend on their size (Eparvier et al., 2013).

Although no specific funding has been envisaged for the CoM, many financial instruments in the EU budget have been available for local authorities to support their activities within the CoM framework. This opportunity has been widely exploited, because according to nearly 60% of respondents, participation in the CoM has been important for better access to EU funding.

The symbolic and legitimising impact of the CoM appears to have been relevant as well, as its signatories have pointed out a high added value of the programme for aspects such as mobilisation of stakeholders, raising awareness among citizens about climate objectives and higher support from regions and enhanced overall visibility of local authorities at the regional, national and EU scales. Additionally, the establishment of multilevel cooperation networks within Member States (municipalities and their territorial coordinators) has been highlighted as an important added value of the CoM related activities. Among others, the experience of Rennes

Métropole in France or Andalusia in Spain has shown that participation in the CoM has contributed to intensify cooperation between local and regional tiers of government in the field of energy and climate policies.

According to the EU Commission (JRC, 2016), the overall experience of the CoM has been successful and the relevance of the CoM as a flexible bottom-up platform for achieving EU2020 Climate and Energy targets has been high, with particular regard to the mitigation commitments related to emissions from energy consumption in the sectors that are widely covered by local jurisdictions. The following measures have been most commonly implemented at the local level:

- Energy management and procurement (district energy networks)
- Building standards and energy certification labelling for new and existing buildings
- Awareness raising and training
- Financial incentives
- Urban planning: local mobility plans defining limited traffic zones, low-emission zones, designated parking services for low-emission vehicles, free parking for cleaner efficient vehicles, integrated ticketing to foster sustainable mobility

A further contribution of municipalities to the reduction of CO_2 emissions and the improvement of energy efficiency have derived from their jurisdiction over local energy and water utilities, public transport and social housing. Although the dynamic of the programme's implementation varied substantially across the EU Member States, the combination of effective urban energy policies and better coordination between national and local governments has been considered crucial for the potential of the urban mitigation of climate change (JRC, 2016).

8.4 Barriers and Obstacles to Participation in the CoM

The implementation of the CoM has been far from unproblematic partly because of the extreme diversity of local conditions that the programme had to tackle, and partly because of the programme's own shortcomings. With regard to the former aspect, the major problematic point concerned the preparation of SE(C)APs and, in particular, the capability of municipalities to align existing methodologies with the EU framework and requirements (Technopolis et al., 2013). In some cases, municipalities have straightforwardly found complementarity between existing local policy tools and the CoM guidance, while others had to put significant effort to align their instruments. In the cases where municipalities have drafted their plans from scratch, a significant level of technical capacities and internal expertise has been required, although by far not always available.

In general, as the CoM implementation show, signatories have taken around 18 months on average to to deliver their SEAPs against 12 months foreseen in the guidance (Eparvier et al., 2013). Furthermore, a considerable variation of timing has been observed even between municipalities of the same country, as it has taken up to 4 years for some municipalities to comply with all requirements.

In addition to methodological discrepancies, the shortage of human and financial resources has widely been mentioned among the main barriers to the smooth preparation of SE(C)APs (72%), followed by absence of reliable data, technical weaknesses and coordination problems. A lack of national and regional support, changes in political context and opposition from civil society have been viewed as much less relevant obstacles for successful participation in the CoM (Eparvier et al., 2013). In addition to the local authorities' opinions and perceptions on the CoM reported above, it has been suggested that the interest to participate in the initiative has varied substantially across the EU, depending on the degree of maturity of municipal sustainable energy policies (Bendlin, 2015).

A lack of access to funding has been mentioned among the main barriers for SE(C)AP implementation by small municipalities, in particular in the Eastern and Southern European countries. Significant financial constraints have seriously hampered the implementation of the CoM, considering that the implementation of actions included in local plans has been mainly covered by municipal budgets. In fact, it has been estimated that the implementation of SEAPs has needed far larger investments than municipalities could afford, considering that 40 billion in investments in energy efficiency and renewable energy sources were set out in 800 action plans examined until 2012 (JRC, 2016). Therefore, the capacity to attract additional public and private funding has been fundamental for cities' successful participation in the programme.

Some funding schemes in EU programmes such as Life+, Horizon 2020 and Intelligent Energy Europe (IEE) as well as EU structural funds (SFs) have targeted to CoM-related actions. However, access to these funding cannot be taken for granted. In the case of EU thematic programmes, funding has been available on competitive basis at the EU level and it is difficult to estimate the exact amount that has been obtained by municipalities for their action plans. The only exception has been the IEE Programme, which has been the main funding source to support several activities relevant for developing and implementing SEAPs. More specifically, this programme has supported a range of activities for promoting energy efficiency and renewable energy across the EU, which were directly connected with and highly relevant for the CoM. The Local Energy Leadership action has been particularly effective in facilitating the implementation of the CoM and although data on the share of funding directly linked to CoM activities are not available, numerous examples of projects supporting CoM activities exist (Eparvier et al., 2013).

The availability of EU Structural Funds (SFs) has been contingent to the political choice of domestic governments in charge of the management of funds at the national and regional levels. In addition to national and regional operational programmes supported by the European Regional Development Fund, other initiatives have been relevant, including Joint European Support for Sustainable Investments in City Areas (JESSICA), Joint Assistance to support Projects in European Regions (JASPERS), INTERREG cooperation programmes, URBACT, Smart Cities and Communities and European Energy Efficiency Facility and the Sustainable Energy Initiative (Eparvier et al., 2013). Among several critical issues that have emerged with regard to the aforementioned funding mechanism, obstacles such as the lack of information on funding opportunities or insufficient appropriateness of financial instruments (i.e., the amount, typology of investments) for specific needs that municipalities had to face when designing or implementing their SEAPs have most often been mentioned. Importantly, the situation has improved during the 2014–2020 programming compared to the 2007–2013 period, as a stronger reference to climate objectives has been included in the EU SF regulations.

Importantly, although the CoM does not offer funding, a limited but relevant leverage effect has been observed on public spending at the local level in the field of energy efficiency/renewable energy investments, with many cities allocating a growing amount of resources to such actions as a result of their participation in the CoM. In this perspective, the involvement of private actors and investments has proved to be highly relevant for the overall success of the CoM, although the data have so far shown that the CoM was not successful in helping the signatories to tie the private actors into the project and to make them reduce their CO_2 production. According to 78% of respondents, participation in the CoM had no effect on the local and regional industrial actors in guiding them to work towards the CoM goals, whereas for 39% such effects could be considered limited (Eparvier et al., 2013). Additional funding provided through capital market, investment and capital banks, foundations and innovative instrument (e.g., crowd funding) has been intensely encouraged by the CoM and an increasing emphasis has been made on the need to assess the long-term outcome of actions in terms of job creation and the promotion of economic sustainability. Language barrier still remains an important issue. The fact that the communication is by and large made in English appears to reduce the CoM effectiveness as well as the ability to effectively communicate towards extended audience.

8.5 The Dynamic Accountability of Multilevel Settings: Strengths and Weaknesses of Flexible Commitment

In addition to the coordination and learning tools described above, the CoM governance architecture has been characterised by a composite multilayered accountability mechanism. First of all, independent of the nature of their commitment, all signatories accept that their membership may be suspended if they fail to submit the required documents (SE(C)APs and monitoring reports) or do not respect the established requirements. Those signatories that have not complied with the requirements are put 'on hold' until they do not provide the required corrections or missing documents.

Before applying for the CoM membership, local authorities willing to sign up must discuss and approve future commitments in their Municipal

Council (or equivalent) thereby ensuring a long-term political support for action. Once the decision to sign up is formally adopted at the local level, the online adhesion procedure must be completed by uploading the duly filled and signed form on the individual page on the Covenant website. A detailed profile is created for all CoM signatories, presenting the data from BEIs, the text of local SE(C)AP, monitoring data and, whenever available, benchmarks to share with other CoM participants. The process of policy convergence was expected to be enhanced through facilitated interaction and transparency that would at the same time enable better comparability, coordination and mutual trust among CoM participants. By making openly available all relevant documents to the CoM community and the public, the programme not only helps keep track of the progress but also contributes to increase the visibility and recognition of efforts taken by its member municipalities to combat climate change, attempting also to strengthen the sense of mutual credibility and fairness among signatories. This system responds to the criteria of the so-called 'dynamic accountability' based on the process of deliberation, which uses the concrete experiences of actors' differing reactions to current problems to generate novel possibilities for consideration rather than buffering decision-makers to make decisions. In such experimentalist governance architectures based on 'deliberative polyarchy' (Sabel and Zeitlin, 2010), the local level units face similar problems, and can learn much from their separate efforts to solve them, even though particular solutions will rarely be generalisable in any straightforward way. The widespread institution of common metrics, peer review and expert evaluation have been supposed to undercut the very notion of technocratic authority associated first of all with the EU Commission monopoly on legislative initiative or implementation monitoring (Sabel and Zeitlin, 2010).

Besides the aforementioned tools aimed at ensuring the individual responsibility of CoM members, a system of multilevel system of support and legitimation has been set out, involving other public authorities and private actors concerned. In fact, it has been expected that if a signatory lacks competences or resources for preparing or implementing its Action Plan, it can rely on a wider political support and technical assistance of national, regional or other territorial bodies that possess the required capacities and means. Therefore, the so-called Covenant

Coordinators, including provinces, regions, ministries, metropolitan cities and other groupings of local authorities, have been invited to join in order to provide signatories of the programme with strategic guidance and technical or financial support. There were 208 Coordinators as of January 2019, most of which come from Italy, Spain, Belgium and France. Evidence has shown that Coordinators have actually offered significant support to the implementation of the CoM, as it has been most successful in regions where coordinators were very active. Coordinators have not only encouraged municipalities to join the CoM but also assisted them in the process of drafting, refining and implementing their plans by helping adapt the CoM methodology to the local policy and institutional conditions. This support was fundamental especially immediately after the establishment of the programme, as no common guidance and methodologies were available yet (Eparvier et al., 2013). Last but not least, Coordinators have helped identify financial opportunities available for the preparation or implementation of SE(C)APs, in particular within the framework of EU direct programmes (e.g., IEE, Life+), for which most CoM member municipalities were too small to apply. Moreover, while often acting as managing authorities for EU SF, coordinators designed specific strands of funding for CoM-related activities in their respective territories. Overall, the assistance provided by coordinators has been highly appreciated by CoM members, being ranked as the second important support tool in the preparation of SEAP (Eparvier et al., 2013).

By the same token, Covenant Supporters representing specialised public and non-profit agencies and networks have been welcomed to participate with a two-fold objective. On the one hand, they could support the commitments of signatories with expert knowledge of the regulatory, legislative and financial framework under which they operate at the respective territorial level. On the other hand, their contribution was considered crucial for mobilising a wider advocacy of CoM activities among local authorities and citizens. The activity of Supporters for capacity building and developing synergies between CoM actions and existing domestic policy instruments has been particularly relevant in the Central and Eastern European countries, where numerous networks that perform activities for energy and environmental protection operate.

Importantly, the CoM guidance on the preparation and implementation of SE(C)AP has called for the engagement of local communities and all relevant stakeholders in the whole process from the very early stage of the preparation of BEIs in order to increase the local ownership of local climate objectives and mobilise resources and capacities required for taking necessary actions. Previous research has highlighted the relevance of these aspects, showing that the active involvement of citizens and stakeholders may increase the public acceptance of climate policies, being also fundamental for their successful implementation (Christoforidis et al., 2013; Larsen and Gunnarsson-Östling, 2009; Lee and Painter, 2015; Pasimeni et al., 2014).

Along with numerous opportunities offered by a multifaceted design of the CoM for developing and upgrading sustainable energy and climate policies at the local level, a number of operational problems and drawbacks have emerged in the implementation of the programme. By adopting an extremely flexible approach, the CoM has aimed to accommodate a vast diversity of policy and regulatory conditions at the local level across the EU. However, this element of potential strength has turned out to be a weakness, as it was hardly possible to meet the very different and even opposite expectations of municipalities from countries where similar domestic instruments existed before compared to those that did not have any regulatory framework for local energy and climate action. It is not surprising that the largest share of CoM members is concentrated in Belgium, Italy, Portugal and Spain, where no domestic schemes for local climate and sustainable energy action existed before, but which could rely on substantial technical and relational potential guaranteed by territorial Coordinators. Instead, less active participation has been observed both in countries with a consolidated experience of TMN for climate (e.g. Germany and France) and in those where neither previous experience nor sufficient resources for developing CoM-related activities were available (e.g. Poland). In the former group, mainly leaders or pioneers have joined the CoM primarily with the purpose of increasing their visibility and developing wide international networks, disregarding additional administrative costs entailed by the whole SE(C)AP procedure. The latter group has been lagging behind because of the lack of technical knowledge, human and financial resources needed for the preparation and implementation of local plans. Evidence exists that CoM learning and coordination resources have not been sufficient to fill in the aforementioned gaps due to a number of shortcomings that are briefly outlined below.

First of all, the operational capacity of the CoMO has been limited due to a significant shortage of human resources. In 2013, its staff was composed of 12 FTE and additional staff, some of which was part-time, while the CoM membership counted 4,638 signatories and it almost doubled by 2018. According to the service contract conditions, the CoMO has been supposed to fulfil tasks such as promotion, communication, support to signatories, including helpdesk and monitoring, and liaison with facilitators. Obviously enough, such small staff could hardly effectively meet communication, information and more specific needs and demands of thousands of members coming from 28 EU and extra-European countries, who often were able to interact only in their mother tongue. Naturally enough, the level of satisfaction of CoM members with the CoMO support has been rather low, with a consistent share of those who either did not use its support at all or assessed it as poor or very poor (Eparvier et al., 2013). Furthermore, it has been stressed that the CoMO was not able to meet the signatories' demand for organisation of information and training events locally, although this activity was extremely important for offering direct assistance to municipalities, especially in countries where technical capacities were missing or weak as, for example, in the Central and Eastern European countries, such as Czech Republic, Hungary, Slovakia and Poland. Furthermore, greater efforts have been required for improving information gathering, editing and dissemination as well as for facilitating exchanges between signatories, experts and stakeholders (Eparvier et al., 2013).

Surprisingly, the partnership with TMN for climate, with which the CoM has been closely connected since its very inception, did not prove to be as effective as expected. These networks have provided the CoM with a wide visibility, helping also in motivating municipalities to join: the biggest share of local authorities has learnt about the CoM because of their participation in other TMNs. It is worth reminding that the consortium underlying the CoM was initially led by Energy Cities (a network composed of around 1000 European local authorities on sustainable energy) and included Eurocities (an advocacy group of 140 large European cities and their 45 partners); Climate Alliance (bringing together 1699 European municipalities committed to climate); the Council of European Municipalities and Regions—CEMR (Europe's biggest association of local and regional authorities, which includes 130,000 local governments and 60 member associations); the European Federation of Agencies and Regions for Energy and the Environment FEDARNE (regional and local organisations which implement, coordinate and facilitate energy and environment policies) (Bendlin, 2015, p. 2). Efforts have been made to develop a tailored approach to CoM activities at the national level by creating a pool of national experts to further decentralise the helpdesk operations that consider domestic legislative, methodological and financial tools (Eparvier et al., 2013). However, despite potential synergies between the CoM and the aforementioned networks in terms of of functions and staff (some members of the CoMO staff divide their time between CoM activities and their 'host' organisations), the value of these links does not appear to have been sufficiently explored, in particular in terms of assistance to CoM members in their routine activities.

Lastly, more effort appears to be needed to improve the functioning and the complementarity of action of the CoMO and the JRC. While the former provides general assistance, the latter has been in charge of technical issues, including the task of replying to signatories' requests. The respective tasks and responsibilities of CoM members do not seem to be clear to them (Eparvier et al., 2013), and there should be better coordination between them.

Conclusions

In summary, a number of important conclusions can be drawn about the potential, weaknesses and strengths of the CoM. Undoubtedly, the CoM has created an important opportunity for reducing CO_2 emissions at the EU scale and beyond by empowering the lowest level of territorial authorities to act collectively in a coordinated manner to fight climate change. Its success in terms of geographic scope and population coverage makes it a unique and powerful instrument of mobilisation and commitment to climate objectives. The policy impact of the CoM has also been substantial albeit uneven. Empirical research shows that for some signatories, the programme has offered a structured way for implementing national regulations, for others it has opened an opportunity to go beyond and lead a transformation process. Instead, in the absence of consolidated domestic policies, signatories have been offered the possibility to design novel policy measures (Croci et al., 2016), although it has not always been exploited because of multiple context barriers, including the lack of experience, knowledge and resources. Thus, significant margins for the programme's improvement exist, as the number of signatories from the countries where the local potential for sustainable energy and climate change is still underexplored have been limited, with the exception of Italy and Spain. To this end, the programme's complementarity to existing EU regulations and domestic policy instruments should be more carefully assessed and improved.

The operational weaknesses reported above, if not appropriately solved, may challenge the CoM's credibility in the long run, thereby undermining the precious potential it retains in terms of territorial mobilisation and the capacity to transfer local knowledge about climate policies. Increasing opportunities of domestic political support and public funding, as well as ensuring the availability of EU financial instruments for promoting CoM activities appear to be crucial for enhancing the programme' success.

The need to proceed in this direction is confirmed by the relevance of the CoM impact in terms of an overall reduction of GHG emissions. A 23% reduction has been reported between the baseline and monitoring years, mainly due to the reduction of emissions from buildings and subsequent lower energy generation levels, more efficient local heat production from district heating networks and an increased share of renewable sources in decentralised local heating production (JRC, 2016). Importantly, the scope of measures implemented by CoM signatories covers several policy sectors entailing wider social and economic impacts at the local level, as addressing climate change requires an overall revision of urban priorities in relation to energy usage in the built environment, but also in transportation, land use planning, waste and water services. The list of policy instruments activated to this end has most often been composed of a variety of measures, including awareness raising and training (26%), urban and transport planning and regulations (18%), grants and subsidies (17%), standards for monitoring and management of energy (12%), codes and regulations in buildings (11%), public procurement (5%) and third party financing (4%). Moreover, the dimension of behavioural change has been considered particularly relevant, as local action plans have promoted communication and awareness campaigns for stakeholders and population with reference to integrated actions in buildings for improving energy efficiency and use of renewables; campaigns for reducing the annual water consumption/waste production; actions for cleaner and efficient vehicles and eco-driving. Evidence has been provided (Heidrich et al., 2016) that such actions go clearly beyond the energy sector, tacking a range of dimensions that are essential for transition to sustainability, such as protection from hydrological risks (flooding, drought); increasing carbon absorption and the creation of fresh air corridors (green urban areas; parks); support and assistance during the summer months for disabled and elderly people (heat wave strategies, early warning systems).

In addition to the aforementioned policy impacts, the implementation of SE(C)APs has resulted in important governance transformations such as mobilisation of public interest in sustainable energy and climate change, the development of internal cross-sectoral coordination within local administrations and an improved communication towards local community as well as to public and private stakeholders.

Overall, the relevance of the multilevel architecture of the CoM for fighting climate change can hardly be overestimated, considering the linkage that local authorities have with citizens and their role in the implementation of European, national and regional public policies. Coordination and complementary of action represent the main challenge for the ultimate effectiveness of the programme. As CoM actions build on and partly overlap with a range of previous policy initiatives and networks, a stronger internal and external coherence is required to maximise its impact. On the former side, more effort is needed to guarantee the coordination between upper-level policies (EU, national and regional) and local initiatives. In fact, previous research (Heidrich et al., 2016; Kern and Bulkeley, 2009) has found that a multi-scale approach to provisioning of plans and strategies from the EU to national and regional levels can be effective in ensuring that cities will develop solid planning for mitigation and adaptation. In the absence of such support, only large pioneering cities will be able to commit to climate objectives and a considerable gap will emerge between them and the rest of smaller local units that lack financial, human and knowledge capacities required for that (Melica et al., 2018).

Hence, the internal compactness of the CoM should be improved based on a careful assessment of actual benefits and costs it has generated for participating local authorities. In this perspective, not only its contribution with regard to the achievement of EU Climate and Energy goals should be better defined, but also the complementarity of action between various EU political and financial instruments should be warranted in order to maximise the overall positive impact. The leverage effect of the CoM with regard to activities performed by local authorities within other transnational programmes and networks needs to be further explored so as to avoid overlaps and strengthen synergies, considering the capacities and potential of cities of different sizes as well as possible barriers related to national contexts. As mentioned, the CoM is far not the only EU initiative for enhancing local involvement in reaching common goals, and its settings and functions still need to be improved in order to comply with the challenge to act as a powerful platform for promoting local institutional and policy innovations for climate. Constant fine-tuning of this programme in the perspective of local authorities' needs and new political ambitions would add a solid argument in favour of the claim that the EU polycentric structure may facilitate policy innovation through 'a dynamic process of multi-level reinforcement' (Schreurs and Tiberghien, 2007) among the different political poles within the context of decentralised governance, thereby supporting the EU's ambition to act as an international policy leader in the area of climate change (Jordan et al., 2012).

References

Bendlin, L. (October 1-2, 2015) The Covenant of Mayors Experience: Lessons for Fostering Local Climate Policy. Paper prepared for the Climate Change and Renewable Energy Policy in the EU and Canada Workshop, Carleton University, Ottawa.

Busch, H. (2015) Linked for action? An analysis of transnational municipal networks in Germany. *International Journal of Urban Sustainable Development*, 7, 213–231. doi:10.1080/19463138.2015.1057144.

Christoforidis, G.C., Chatzisavvas, K.C., Lazarou, S. and Parisses, C. (2013) Covenant of mayors initiative-Public perception issues and barriers in Greece. *Energy Policy*, 60, 643–655.

Croci, E. et al. (2016) Urban CO2 mitigation strategies under the Covenant of Mayors: An assessment of 124 European cities. *Journal of Cleaner Production*, 169, 161–177.

Eparvier, P. et al. (2013) Mid-term evaluation of the Covenant of Mayors, Final Report, available at www.technopolis-group.com

European Commission (2007) *An Energy Policy for Europe-COM (2007) 1-and Limiting Global Climate Change to 2 Degrees Celsius–The Way Ahead for 2020 and Beyond-COM (2007) 2*, available at: http://ec.europa.eu/transparency/regdoc/rep/2/2008/EN/2-2008-85-EN-1-0.Pdf

European Commission (2015) *Communication to the European Parliament, the Council, the European Economic and Social Committee, the Committee of the Regions and the European Investment Bank. A Framework Strategy for a Resilient Energy Union with a Forward-Looking Climate Change Policy. COM(2015)080 final*, available at https://eur-lex.europa.eu/

European Commission (2016a) *Communication to the European Parliament, the Council, the European Economic and Social Committee, the Committee of the Regions and the European Investment Bank. An EU Strategy on Heating and Cooling. COM(2016)51 Final*, available at: https://ec.europa.eu/energy/sites/ener/files/documents/1_EN_ACT_part1_v14.pdf

European Commission (2016b) *Communication to the European Parliament, the Council, the European Economic and Social Committee, the Committee of the Regions and the European Investment Bank. A European Strategy for Low-Emission Mobility. COM(2016)501 Final*, available at: https://ec.europa.eu/transparency/regdoc/rep/1/2016/EN/1-2016-501-EN-F1-1.PDF

European Union (2009) Decision 406/2009/EC of the European Parliament and of the Council of 23 April 2009 on the effort of Member States to reduce their greenhouse gas emissions to meet the Community's greenhouse gas emission reduction commitments up to 2020. (June 5, 2009) *Official Journal of the European Communities*, OJL 140, 136–148.

European Union (2003) Directive 2003/87/EC of the European Parliament and of the Council of 13 October 2003 establishing a scheme for greenhouse gas emission allowance trading within the Community and amending Council Directive 96/61/EC. Official Journal of the European Communities. (October 25, 2003) *Official Journal of the European Communities*, OJL 275/32, 32–46.

European Union (2012) Directive 2012/27/EU of the European Parliament and of the Council of 25 October 2012 on energy efficiency, amending Directives 2009/125/EC and 2010/30/EU and repealing Directives 2004/8/EC and 2006/32/EC. (November 14, 2012) *Official Journal of the European Communities*, OJL 315, 1–56.

Heidrich, O. et al. (2016) National climate policies across Europe and their impacts on cities strategies. *Journal of Environmental Management*, 16(8), 36–45.

Heyvaert, V. (2013) What's in a name? The Covenant of Mayors as transnational environmental regulation. *Review of European Community and International Law*, 22(1), 78–90.

Joint Research Centre (JRC) (2016) *Covenant of mayors: Greenhouse gas emissions achievements and projections. Science for Policy Report,* available at the website: https://e3p.jrc.ec.europa.eu/sites/default/files/documents/publications/jrc103316_com_achievements_and_projections_online.pdf

Jordan, A. et al. (2012) Understanding the paradoxes of multi-level governing: Climate change policy in the European Union. *Global Environmental Politics*, 12(2), 43–65.

Jordan, A. et al. (2018) *Governing Climate Change. Polycentricity in Action?* Cambridge University Press, Cambridge.

Jordan, A.K. et al. (2015) Emergence of polycentric climate governance and its future prospects. *Nature Climate Change*, 5, 977–982.

Kern, K. (June 23–26, 2010) *Climate Governance in the EU Multi-Level System. The Role of Cities,* Paper presented at the Fifth Pan-European Conference on EU Politics, Porto.

Kern, K. and Bulkeley H. (2009) Cities, Europeanization and multi-level governance: Governing climate change through transnational municipal networks. *Journal of Common Market Studies*, 47(2), 309–332.

Larsen, K. and Gunnarsson-Östling, U. (2009) Climate change scenarios and citizen-participation: Mitigation and adaptation perspectives in constructing sustainable futures. *Habitat International*, 33, 260–266.

Lee, T. and Painter, M. (2015) Urban climate comprehensive local climate policy: The role of urban governance. *Urban Climate*, 14, 566–577.

Melica G. et al. (2018) Multilevel governance of sustainable energy policies: The role of regions and provinces to support the participation of small local authorities in the Covenant of Mayors. *Sustainable Cities and Society*, 39, 729–739.

Pasimeni, M.R. et al. (2014) Scales, strategies and actions for effective energy planning: A review. *Energy Policy*, 65, 165–174.

Sabel, C.F. and Zeitlin J. (eds.) (2010) *Experimentalist Governance in the European Union: Towards a New Architecture*. Oxford, Oxford University Press.

Schreurs, M.A. and Tiberghien Y. (2007) Multi-level reinforcement: Explaining European Union leadership in climate change mitigation. *Global Environmental Politics*, 7(4), 19–46.

Index

CPSIA information can be obtained
at www.ICGtesting.com
Printed in the USA
BVHW042338240720
584036BV00011B/26

9 781786 348142

THE TRANSFORMATION OF THE EUROPEAN UNION

The Impact of Climate Change in European Policies

THE TRANSFORMATION OF THE EUROPEAN UNION

The Impact of Climate Change in European Policies

Editor

Xira Ruiz-Campillo

Complutense University of Madrid, Spain

World Scientific

NEW JERSEY • LONDON • SINGAPORE • BEIJING • SHANGHAI • HONG KONG • TAIPEI • CHENNAI • TOKYO

Published by

World Scientific Publishing Europe Ltd.
57 Shelton Street, Covent Garden, London WC2H 9HE
Head office: 5 Toh Tuck Link, Singapore 596224
USA office: 27 Warren Street, Suite 401-402, Hackensack, NJ 07601

Library of Congress Cataloging-in-Publication Data
Names: Ruiz-Campillo, Xira, editor.
Title: The transformation of the European Union : the impact of climate change in European policies / editor, Xira Ruiz-Campillo, Complutense University of Madrid, Spain.
Description: London ; Hackensack, NJ : World Scientific Publishing Europe Ltd., [2020] | Includes bibliographical references and index.
Identifiers: LCCN 2019047660 | ISBN 9781786348142 (hardcover) | ISBN 9781786348159 (ebook) | ISBN 9781786348166 (ebook other)
Subjects: LCSH: Environmental policy--European Union countries. | European Union countries--Environmental conditions. | Climatic changes--Political aspects--European Union countries.
Classification: LCC HC240.9.E5 T73 2020 | DDC 363.738/74561094--dc23
LC record available at https://lccn.loc.gov/2019047660

Library Cataloguing-in-Publication Data
A catalogue record for this book is available from the British Library.

For any available supplementary material, please visit
https://www.worldscientific.com/worldscibooks/10.1142/Q0241#t=suppl

Typeset by Diacritech Technologies Pvt. Ltd.
Chennai - 600106, India

To every single person working for a more sustainable and better world.